UJFG   NATO ASI

LES HOUCHES

Session LXV
1996

DE LA CELLULE AU CERVEAU

Le Cytosquelette
Communication Intra-et Inter-Cellulaire
Le Système Nerveux Central

FROM CELL TO BRAIN:

The Cytoskeleton
Intra- and Inter-Cellular Communication
The Central Nervous System

CONFÉRENCIERS

F. Bourrat
D. Duboule
D. Engelman
A. Goldbeter
B. Goud
D. Job
B. Knaupp
R. Kelly
R. McIntosh
R. Margolis
O. Pourquié
D. Sabatini
J.-L. Sikarov
A. Triller
P. Vernier
R. Wade

UJFG   NATO ASI

LES HOUCHES

SESSION LXV

8 Juillet – 26 Juillet 1996

DE LA CELLULE AU CERVEAU:
Le Cytosquelette
Communication Intra-et Inter-Cellulaire
Le Système Nerveux Central

FROM CELL TO BRAIN:
The Cytoskeleton
Intra- and Inter-Cellular Communication
The Central Nervous System

édité par
G. ZACCAI, J. MASSOULIÉ
et F. DAVID

1998

ELSEVIER
Amsterdam – Lausanne – New York – Oxford – Shannon – Singapore – Tokyo

ELSEVIER SCIENCE B.V.
Sara Burgerhartstraat 25
P.O. Box 211, 1000 AE Amsterdam, The Netherlands

First edition 1998

Library of Congress Cataloging in Publication Data
A catalog record from the Library of Congress has been applied for.

ISBN: 0 444 50086 3

⊗ The paper used in this publication meets the requirements of ANSI/NISO Z39.48-1992 (Permanence of Paper)

Printed in The Netherlands

LES HOUCHES
ÉCOLE D'ÉTÉ DE PHYSIQUE THÉORIQUE

SERVICE INTER-UNIVERSITAIRE COMMUN
À L'UNIVERSITÉ JOSEPH FOURIER DE GRENOBLE
ET À L'INSTITUT NATIONAL POLYTECHNIQUE DE GRENOBLE,
SUBVENTIONNÉ PAR LE CENTRE NATIONAL DE LA RECHERCHE
SCIENTIFIQUE ET LE COMMISSARIAT À L'ÉNERGIE ATOMIQUE

SESSION LXV
INSTITUT D'ÉTUDES AVANCÉES DE L'OTAN
NATO ADVANCED STUDY INSTITUTE
8 JUILLET - 26 JUILLET 1996

Directeurs scientifiques de la session: Guiseppe Zaccai, *Institut de Biologie Structurale,
41 avenue des Martyrs, 38027 Grenoble Cedex 01, France et* Jean Massoulié, *Ecole
Normale Supérieure, Neurobiologie Cellulaire et Moléculaire, 46 rue d'Ulm, 75230
Paris Cedex 05, France*

SESSIONS PRÉCÉDENTES

XXXII	1979	Cosmologie physique. Physical cosmology (North-Holland)*
XXXIII	1979	Membranes et communication intercellulaire. Membranes and intercellular communication (North-Holland)*
XXXIV	1980	Interaction laser-plasma. Laser-plasma interaction (North-Holland)*
XXXV	1980	Physique des défauts. Physics of defects (North-Holland)*
XXXVI	1981	Comportement chaotique des systèmes déterministes. Chaotic behaviour of deterministic systems (North-Holland)*
XXXVII	1981	Théories de jauge en physique des hautes énergies. Gauge theories in high energy physics (North-Holland)*
XXXVIII	1982	Tendances actuelles en physique atomique. New trends in atomic physics (North-Holland)*
XXXIX	1982	Développements récents en théorie des champs et mécanique statistique. Recent advances in field theory and statistical mechanics (North-Holland)*
XL	1983	Relativité, groupes et topologie II. Relativity, groups and topology II (North-Holland)*
XLI	1983	Naissance et enfance des étoiles. Birth and infancy of stars (North-Holland)*
XLII	1984	Aspects cellulaires et moléculaires de la biologie du développement. Cellular and molecular aspects of developmental biology (North-Holland)*
XLIII	1984	Phénomènes critiques, systèmes aléatoires, théories de jauge. Critical phenomena, random systems, gauge theories (North-Holland)*
XLIV	1985	Architecture des interactions fondamentales à courte distance. Architecture of fundamental interactions at short distances (North-Holland)*
XLV	1985	Traitement du signal. Signal processing (North-Holland)*
XLVI	1986	Le hasard et la matière. Chance and matter (North-Holland)*
XLVII	1987	Dynamique des fluides astrophysiques. Astrophysical fluid dynamics · (North-Holland)*
XLVIII	1988	Liquides en interfaces. Liquids at interfaces (North-Holland)*
XLIX	1988	Champs, cordes et phénomènes critiques. Fields, strings and critical phenomena (North-Holland)*
L	1988	Tomographie océanographique et géophysique. Oceanographic and geophysical tomography (North-Holland)*
LI	1989	Liquides, cristallisation et transition vitreuse. Liquids, freezing and glass transition (North-Holland)*
LII	1989	Chaos et physique quantique. Chaos and quantum physics (North-Holland)*
LIII	1990	Systèmes fondamentaux en optique quantique. Fundamental systems in quantum optics (North-Holland)*
LIV	1990	Supernovae (North-Holland)*
LVI	1991	Fermions en forte interaction et supraconductivité à haute température. Strongly interacting fermions and high t_c superconductors (North-Holland)*
LVII	1992	Gravitation et quantifications. Gravitation and quantizations (North-Holland)*
LVIII	1992	Les progrès du traitement des images. Progress in picture processing (North-Holland)*
LIX	1993	Mécanique des fluides numérique. Computational fluid dynamics (North-Holland)*
LX	1993	Cosmologie et structure à grande échelle. Cosmology and large scale structure (North-Holland)*
LXI	1994	Physique quantique mésoscopique. Mesoscopic quantum physics (North-Holland)*
LXII	1995	Géometries fluctuantes en mécanique statistique et en théorie des champs. Fluctuating geometries in statistical mechanics and field theory (North-Holland)*
LXIV	1995	Symétries quantiques. Quantum symmetries. (North-Holland)*

---

* Sessions ayant reçu l'appui du Comité Scientifique de l'OTAN

LECTURERS

Bourrat, Franck   Laboratoire de Génétique des Poissons, INRIA Centre de Recherche, Domaine de Vilvert, 78352 Jouy en Josas Cedex, France. Tel.: 33 1 34 65 23 91/92; Fax: 33 1 34 65 23 90/27 04.

Duboule, Denis   Dept. of Zoology, Sciences 111, 30 Quai Ernest-Ansermet, 1211 Geneve 4, Switzerland. Tel.: 41 22 702 67 71; Fax: 41 22 702 67 95.

Goud, Bruno   Institut Curie, rue Lhomond, 75230 Paris Cedex 5, France. Tel.: 33 1 42 34 63 98

Job, Didier   DBMS/CS, CEA/Grenoble, 17 Avenue des Martyrs, 38054 Grenoble Cedex 9, France. Tel.: 33 4 76 88 38 01.

McIntosh, Richard   Dept. of Molecular, Cellular and Developmental Biology, University of Colorado, Boulder, CO 80309-0347, USA. Tel.: 1 303 492 8533; Fax: 1 303 492 7744; E-mail: dick@beagle.colorado.edu.

Margolis, Robert   IBS, 41 Avenue des Martyrs, 38027 Grenoble Cedex 1, France. Tel.: 33 76 88 96 16; Fax: 33 76 88 54 94; E-mail: margolis@ibs.fr.

Pourquié, Olivier   Institut d'Embryologie, 49 bis Avenue de la Belle Gabrielle, 94736 Nogent sur Marne, France. Tel.: 33 1 45 14 15 15; Fax: 33 1 48 73 43 77; E-mail: pourquie@infobiogen.fr.

Sabatini, David   Dept. of Cell Biology, NYU Medical Center, 550 First Avenue, New York, NY 10016, USA

Vernier, Philippe   CNRS, Bât. 32–33 Avenue de la Terrasse, 91198 Gif sur Yvette Cedex, France. Tel.: 33 1 69 82 34 39; Fax: 33 1 69 07 05 38; E-mail: vernier@iaf.cnrs-gif.fr

Wade, Richard   IBS, 41 Avenue des Martyrs, 38027 Grenoble Cedex 1, France Tel.: 33 4 76 88 40 24; Fax: 33 4 76 88 54 94; E-mail: wade@ibs.fr.

SEMINAR SPEAKERS

Engelman, Donald   Dept. of Molecular Biophysics and Biochemistry, YALE University, 266 Whitney Avenue, 420 Bass Center, New Haven, CT 06520, USA. Tel.: 1 203 432 5601; Fax: 1 203 432 6381.

Goldbeter, Albert   Unité de Chronobiologie Théorique, Université Libre de Bruxelles, Campus Plaine C.P. 231, 1050 Brussels, Belgium. Tel.: 32 2 650 57 72; Fax: 32 2 650 57 67. E-mail: agoldbet@ulb.ac.be.

Kaupp, Benjamin   Forschungszentrum Jülich (KFA) IBI, 25425 Julich, Germany. Tel.: 49 24 61 61 40 41; Fax: 49 24 61 61 42 16.

Kelly, Regis   Hormone Research Institute, University of California, San Francisco, CA 94743-05434, USA. Tel.: 1 415 476 4095; Fax: 1 415 731 3612; E-mail : kelly@cgl.ucsf.edu.

Sikorav, Jean-Louis   SBGM CEA Saclay, Bât. 142, 91190 Gif sur Yvette Cedex, France. Tel.: 33 1 69 08 66 43; Fax: 33 1 69 08 47 12.

Triller, Antoine   ENS, 46 Rue d'Ulm, 75230 Paris Cedex 05, France. Tel.: 33 1 44 32 35 47; Fax: 33 1 44 32 36 54; E-mail: triller@wotan.ens.fr.

PARTICIPANTS

Asipauskas, Marius   Dept. of Physics, University of Notre Dame, Notre Dame, IN 46556, USA. Tel.: 1 219 631 9351; Fax: 1 219 631 5952; E-mail: Marius.Asipauskas.1@nd.edu.

Bar-ziv, Roy   Dept. of Physics of Complex Systems, Weizmann Institute of Science, Rehovot, Israel 76100.

Bartoli, Marc   LNCF/CNRS UPR 9013 31, Chemin J. Aiguier, 13042 Marseille Cedex 20, France. Tel.: 33 4 91 16 41 66; Fax: 33 4 91 71 89 14; E-mail: bartoli@lnf.cnrs-mrs.fr.

Bignami, Fabrizia   Dep. Cell. Biol. and Development, P. le Aldo Moro 5, 00185 Rome, Italy. Tel.: 39 6 499 12131; Fax: 39 6 499 12351.

Bourdieu, Laurent   CNRS, Inst. de Physique, Laboratoire LUDFC, 3 Rue de l'Université, 67084 Strasbourg, France. Tel.: 33 3 88 35 80 65; Fax: 33 3 88 35 80 99; E-mail: bourdieu@fresnel.u-strasbg.fr.

Bozdogan, Ömer   Kafkas Univ. Veteriner Fakültesi, Fizyoloji Bilim Dali, Kars, Turkey. Tel.: 90 474 2123091; Fax: 90 474 2237958.

Cave, Adrien   Centre de Biochimie Structurale, Faculté de Pharmacie, 15 Avenue Charles Flahault, 34060 Montpellier Cedex, France. Tel.: 33 4 67 04 34 30; Fax: 33 4 67 52 96 23.

Celebi, Gurbuz   Ege Univ. Tip Fakultesi, Biyofizik AD, 35100 Bornova Izmir, Turkey. Tel.: 90 232 388 2868; Fax: 90 232 388 2868; E-mail: gcelebi@bornova.ege.edu.tr.

Csahok, Zoltan   Dept. of Atomic Physics, Eötvos Univ., 1088 Budapest Puskin U.5-7, Hungary. Tel.: 36 1 266 79 02; Fax: 36 1 266 02 06; E-mail: csahok@hercules.elte.hu.

Czirok, Andras   Dept. of Atomic Physics, Eötvös Univ., 1088 Budapest Puskin U.5-7, Hungary. Tel.: 36 1 266 79 02; Fax: 36 1 266 02 06; E-mail: czirok@hercules.elte.hu.

Derenyi, Imre   Dept. of Atomic Physics, Eötvos Univ., 1088 Budapest Puskin U.5-7, Hungary. Tel.: 36 1 266 79 02; Fax: 36 1 266 02 06; E-mail: derenyi@hercules.elte.hu.

Dutton, Anna   Dept. of Biological Sciences, Univ. of Durham, Science Laboratories, South Rd, Durham DH I 3LE, UK. Tel.: 44 191 374 2000; Fax: 44 191 374 2417.

Dzeja, Claudia   Inst. of Biological Information Processing, KFA Research Centre Jülich, 52425 Julich, Germany. Tel.: 49 2461 61 4753; Fax: 49 2461 61 4216; E-mail: c.dzeja@kfa-julich.de.

Fleury, Vincent   Laboratoire de Physique de la Matière Condensée, Ecole Polytechnique, 91128 Palaiseau, France. Tel.: 33 1 69 33 42 87; Fax: 33 1 69 33 30 04; E-mail: vincent.fleury@polytechnique.fr.

Fourcade, Bertrand   Maison des Magistères Jean Perrin, CNRS, 25 Avenue des Martyrs, BP 116, 38042 Grenoble Cedex, France. Tel.: 33 4 76 88 79 84; Fax: 33 4 76 88 79 81; E-mail: fourcade@ujf-grenoble.fr

Gerbal, Fabien   Inst. Curie, Section de Recherche URA 1379, 11 Rue Pierre et Marie Curie, 75231 Paris Cedex 05, France. Tel.: 33 1 42 34 67 69; Fax: 33 1 40 51 06 36; E-mail: gerbal@curie.fr.

Gjerset, Ruth Anne   Sidney Kimmer Cancer Center, 3099 Science Park Road, San Diego, CA 92122, USA. Tel.: 1 619 450 5990 poste 232; Fax: 1 619 450 3251; E-mail: ruthgj@aol.com.

Grob, Patricia   Inst. de Biologie Structurale, 41 Avenue des Martyrs, 38027 Grenoble Cedex 1, France. Tel.: 33 4 76 88 95 66; Fax: 33 4 76 88 54 94; E-mail: grob@ibs.fr.

Gülöksüz, Asli   Middle East Technical Univ., Dept. of Electrical and Electronics Engineering, 06531 Ankara, Turkey. Tel.: 90 312 210 4425; Fax: 90 312 210 1261; E-mail: guloksuz@rorqual.cc.metu.edu.tr.

Hauss, Thomas   Hahn–Meitner Institut Berlin, Glienicker Str. 100, D 14109 Berlin, Germany. Tel.: 49 30 80622071; Fax: 49 30 80622999; E-mail: hauss@hmi.de.

Hebraud, Pascal   LUDFC, 3 Rue de l'Université, 67000 Strasbourg, France. Tel.: 33 3 88 35 81 27; Fax: 33 3 88 35 80 99; E-mail: hebraud@fresnel. u-strasbg.fr.

Heidenreich, Michael   Sektion Kernspinresonanz Spektroskopie, Univ. Ulm, A. Einstein Allee 11, 89069 Ulm, Germany. Tel.: 49 731 502 3149; Fax: 49 731 502 3150; E-mail: michael.heidenreich@physik.uni-ulm.de.

Jacrot, Bernard   Rue de la Place, 84160 Cucuron, France.

Kern, Norbert   Laboratoire d'Expérimentation Numérique, Maison des Magistères, BP 116, 38042 Grenoble Cedex, France. Tel.: 33 4 76 88 79 84; Fax: 33 4 76 88 79 81.

Khrustova, Natalia   Int. of Biochemical Physics, Russian Academy of Science, ul. Kosygina 4, Moscow 117977, Russia. Tel.: 7 095 939 73 51; Fax: 7 095 135 44 06; E-mail: chembio@glas.apc.org.

Lafanechere, Laurence   Laboratoire du Cytosquelette, INSERM U366, CEA/ Grenoble, 17 Rue des Martyrs, 38054 Grenoble Cedex 9, France. Tel.: 33 4

76 88 38 01; Fax: 33 4 76 88 50 57; E-mail: lafanechere@seine.ceng.cea.fr.

Le Bouar, Tugdual   LPS Dept. de Physique, Ecole Normale Supérieure, 24 Rue Lhomond, 75231 Paris Cedex 05, France. Tel.: 33 1 44 32 32 53; Fax: 33 1 44 32 34 33; E-mail :lebouar@physique.ens.fr.

Limozin, Laurent   IRPHE Biophysique, Case 252 Centre St. Jerome Av. Esc. Normandie Niemen, 13397 Marseille, France. Tel: 33 4 91 28 81 21 Fax: 33 4 91 63 52 61 E-mail: limo@lrc.univ-mrs.fr

Messina, Samantha   Dip. di Medicina Sperimentale e Scienze Biochimiche, Cattedra de Fisiologia Umana, Piano I, Edificio F, Settore sud, Via di Tor Vergata 135, 00133 Rome, Italy. Tel.: 39 6 72 59 64 28; Fax: 39 6 20 42 72 92; E-mail: messina@tovvx1@ccd.utovrm.it.

Noireaux, Vincent   Laboratoire PSI, Section de Recherche, 11 Rue Pierre et Marie Curie, 75231 Paris Cedex 05, France. Tel.: 33 1 33 92 60 14; Fax: 33 1 40 51 06 36.

Palacz, Krysztof   Solid State Theory Division, Institute of Physics, Adam Mickiewicz University, Matejki 48/49, Poznan, Poland. E-mail: kappa@math.amu.edu.pl.

Papadopoulos, Georgios   University of Thessaly, School of Medicine, Larisa, Greece. E.mai: gpap@olymp.ccf.aufh.gr.

Parello, Joseph   UPRES. A, CNRS 5074, Chimie Biomoleculaire et Interactions Biologiques, Faculté de Pharmacie, Bât K, 15 Avenue Charles Flohault, 34060 Montpellier Cedex, France; Cancer Research Center, 10901 North Torrey Pines Rd, La Jolla, CA, USA. Tel.: 33 4 67 52 23 01; Fax: 33 4 67 04 21 40; E-mail: joseph@ljcrf.edu.

Pebay Peroula, Eva   Inst. de Biologie Structurale, 41 Avenue des Martyrs, 38027 Grenoble Cedex 1, France. Tel.: 33 4 76 88 95 83; Fax: 33 4 76 88 54 94; E-mail: pebay@ibs.fr.

Pelce, Pierre   IRPHE/Biophysique, Univ. de Provence St Jerome, Case 252, 13397 Marseille Cedex 20, France. Tel.: 33 4 91 28 81 25; Fax: 33 4 91 63 52 61; E-mail: pelce@lrc.univ-mrs.fr.

Raghavachari, Sidhar   Dept. of Physics, Univ. of Notre Dame, Notre Dame, IN 46556, USA. Tel.: 1 219 631 9351; Fax: 1 219 631 5952; E-mail: sraghava@couperin.phys.nd.edu.

Rascon Diaz, Carlos   Dpto. Fisica Teorica de la Materia Condensada, Facultad de Ciencas, CV 28049 Madrid, Spain. Tel.: 34 1 397 86 46; Fax: 34 1 397 49 50; E-mail: rascon@fluid1.fmc.uam.es.

Réat, Valérie   Inst. de Biologie Structurale, 41 Avenue des Martyrs, 38027 Grenoble Cedex 1, France. E-mail: reat@ibs.fr.

Russ, William   Yale Univ., Dept. of Molecular Biophysics and Biochemistry, 266 Whitney Ave., Rm 429 Bass, New Haven, CT 06520, USA. Tel.: 1 203 452 5602; Fax: 1 203 432 5175; E-mail: russ@paradigm.csb.yale.edu.

Shih, William   Dept. of Biochemistry, Beckman Center, Room B407, Stanford Univ., School of Medicine, Stanford, CA 94305-5307, USA. Tel.: 1 415 723 6503; Fax: 1 415 725 6044; E-mail: willshih@leland.stanford.edu.

Spiridonov, Vyacheslav   Inst. for Nuclear Research, Russian Academy of Sciences, 60th October Anniversarypr. 7a, Moscow 117312, Russia. E-mail: spiridonov@lps.umontreal.ca.

Stuhrmann, Sigrid   Groupe GKSS, Forschungzentrum at the Hamburger Synchotron Laboratory Hasylab, Notkestr. 85, 22607 Hamburg, Germany. Tel.: 49 40 89 98 26 76; Fax: 49 40 89 98 27 87.

Uzdensky, Anatoly   Dept. of Biophysics, Inst. of Neurocybernetics, Rostov Univ., Stachki Av., 194/1, Rostov on Don 344090, Russia. Tel.: 7 (8632) 280 577; Fax: 7 (8632) 280 367; E-mail: uzd@krinc.rnd.runnet.ru.

Valiron, Odile   Inst. de Biologie Structurale, 41 Avenue des Martyrs, 38027 Grenoble, France. Tel.: 33 4 76 88 59 55; Fax: 33 4 76 88 50 57.

Vidybida, Alexander   Bogolyubov Inst. for Theoretical Physics, Metrologichna str. 14-B, 252143 Kiev, Ukraine. Tel: 380 44 2669468; Fax: 380 44 2665998; E-mail: vidybida@gluk.apc.org.

Weik, Martin   Inst. de Biologie Structurale, 41 Avenue des Martyrs, 38027 Grenoble Cedex 1, France. Tel.: 33 4 76 88 47 41; Fax: 33 4 76 88 54 94; E-mail: weik@ibs.fr.

Willbrand, Karen   Max Planck Institut für Kolloid und Grenzflaechenforschung, Kantstr. 55, 14513 Teltow, Germany. E-mail: kawi@mpikg-teltow.mpg.de.

Yetkin, Yalçin   Science Faculty of Atatürk Univ., Dept. of Biology, Inst. of Physiology, 25240 Erzurum, Turkey. Tel.: 90 (442) 233 80 28; Fax: 90 (442) 233 10 62.

Zaccai, Nathan   Christ Church, University of Oxford, Oxford, UK.

PRÉFACE

L'Ecole d'Eté des Houches de juillet 1996 a été consacrée à une présentation de différents aspects de la biologie cellulaire, pour les étudiants d'autres disciplines, en particulier des physiciens. Mais qu'est-ce que la biologie cellulaire? Et, si nous nous posons de telles questions, qu'est-ce que la physique? Quelqu'un a dit que la physique est ce que font les physiciens. Le corollaire qui vient immédiatement à l'esprit, c'est que les physiciens ne font pas de biologie. Par conséquent, des physiciens sont venus aux Houches pour apprendre ce qu'est la biologie cellulaire, au contact de scientifiques de cette discipline. On peut en effet caractériser la biologie cellulaire comme l'étude des cellules et de leur organisation: le niveau minimal et universel de la Vie.

En plus de la satisfaction culturelle d'entendre des experts leur parler de la "Vie", les physiciens avaient deux types de questions d'intérêt professionnel: 1) pouvaient-ils apporter une contribution personnelle à la biologie et à la compréhension des fonctions cellulaires? 2) pouvaient-ils identifier des problèmes de physique intéressants à partir de l'étude des cellules? Apparemment ces questions ne diffèrent que par la motivation qui les sous-tend, mais elles correspondent en fait à des approches tout-à-fait différentes. Ceci est bien illustré dans le cas des membranes biologiques. On a soupçonné dès 1920 qu'une bicouche de lipides constituait la barrière de perméabilité principale des membranes de cellules. Quarante ans plus tard, des études physico-chimiques effectuées sur des lipides extraits de membranes ont pu révéler la richesse de leurs modes d'organisations polymorphes. Pour comprendre leur fonction biologique, il est certainement intéressant de caractériser les propriétés des lipides et leur capacité à former des bicouches. Cependant, ces études ont aussi conduit à d'importants progrès dans la chimie physique des molécules amphiphiles et des surfactants, servant de base à des développements de recherche fondamentale et appliquée qui n'ont que très peu ou pas du tout de rapport avec la biologie.

Il existe une relation bien établie entre la biochimie moléculaire et la chimie physique. Par suite de l'hypothèse de base que la structure détermine la fonction,

Préface

la biologie moderne admet que le rôle biologique d'une macromolécule, par exemple une protéine ou un acide nucléique, dépend de ses propriétés physico-chimiques. L'approche physico-chimique entraîne une exigence fondamentale pour l'analyse biochimique ou structurale au niveau moléculaire: le système étudié (par exemple une protéine déterminée) doit être bien défini et isolé (un biochimiste dirait "pur") et fonctionnellement actif. A priori, on pourrait penser, par conséquent, que ces conditions représentent une exigence absolue pour l'analyse de propriétés biologiques en termes de modèles moléculaires. Cependant, nous verrons que ce n'est pas le cas en biologie cellulaire.

La biologie cellulaire est la science des processus cellulaires. Ceux-ci sont habituellement étudiés en tant que flux moléculaires et structures dynamiques intracellulaires génétiquement contrôlés. Une cellule n'est pas un système complexe statique. Des processus actifs, dynamiques maintiennent un état stationnaire ou déterminent des variations organisées, telles que celles qui se produisent pendant le développement embryonnaire ou l'adaptation à l'environnement. L'exigence de pureté, qui est essentielle pour une étude biochimique et structurale au niveau moléculaire, ne peut pas être maintenue de la même façon en biologie cellulaire parce que tous les acteurs sur la scène cellulaire sont loin d'être connus, et parce que chacun joue son rôle en relation étroite avec les autres. Un enjeu important, dans beaucoup d'expériences, est en fait de déterminer quelles macromolécules sont impliquées dans un processus donné. Si la physique doit intervenir dans ces études, ce devrait être d'une façon très différente de la biologie structurale et de la biochimie, qui se focalisent sur l'analyse de molécules isolées.

La physique apporte une contribution importante, essentielle même, à la biologie dans le domaine de l'instrumentation et des méthodes, et de fait les progrès de la biologie cellulaire sont très largement dus à de nouvelles approches physiques, telles que la microscopie confocale, les méthodes de patch-clamp et les pinces optiques. La question de l'instrumentation et des méthodes a cependant été discutée dans différentes réunions de physiciens et biologistes et n'a pas été traitée en détail au cours de cette école d'été des Houches, qui a été plutôt consacrée à quelques problèmes centraux posés par l'organisation et le fonctionnement cellulaires.

Les biologistes et les physiciens possèdent la caractéristique commune de s'intéresser à des gammes très étendues de niveaux hiérarchiques d'organisation. Le domaine de la biologie s'étend depuis les dimensions spatiales caractéristiques des réactions chimiques, jusqu'à celles des organismes complexes, et finalement à celles de leurs interactions avec l'environnement (écologie), ce qui représente plus de 10 ordres de grandeur, de l'Å au km, et ceci inclut de petites molécules (acides aminés, nucléotides, lipides, sucres, . . .), des macromolécules (ribosomes,

xviii

complexes multienzymatiques), des systèmes intégrés (réticulum endoplasmique, appareil de Golgi, endosomes, vésicules, synapses, ...) jusqu'aux cellules, aux organismes multicellulaires et aux écosystèmes. La biologie est également concernée par un vaste domaine d'échelles de temps, depuis la femtoseconde pour les transitions électroniques, jusqu'aux centaines de millions d'années de l'évolution. Quant à la physique, elle couvre évidemment la totalité des échelles de temps, des dimensions et des degrés d'organisation, depuis les particules élémentaires jusqu'à l'univers entier. Elle s'intéresse à des propriétés fondamentales de la matière, qui s'expriment et s'observent au mieux dans des systèmes simples ou dans les comportements collectifs de systèmes ordonnés. Au niveau moléculaire, de la matière condensée, par exemple, les physiciens sont à l'aise lorsqu'ils analysent les propriétés des cristaux, dans lesquels un même motif atomique est répété de façon régulière et indéfinie, ou celles d'homopolymères placés dans des conditions thermodynamiques bien déterminées. Pour un biologiste, par contre, un cristal n'est qu'un moyen de déterminer au niveau atomique la structure de molécules complexes, telles que des protéines ou des acides nucléiques, afin de progresser dans la compréhension de leur fonction.

Les physiciens sont déroutés par la complexité des systèmes biologiques. La biologie les confronte au défi de simplifier la description d'un système, tout en conservant ses caractéristiques fonctionnelles. Ceci est difficile, même dans le cadre de l'approche réductionniste de la biologie moléculaire: par exemple, chaque protéine est un hétéro-polymère dont les propriétés structurales complexes sont reliées de façon particulière à sa fonction.

En biologie, il existe un flux d'information de l'ADN à l'ARN et aux protéines, suivant le schéma: **ADN** ↔ **ADN** ↔ **ARN** → **protéine**. Le processus de reproduction de l'ADN à partir de lui-même est appelé réplication; la production d'ARN d'après l'ADN est appelée transcription; la production en retour d'ADN à partir d'ARN est la transcription réverse (elle se produit pour certains virus dont le matériel génétique, constitué d'ARN, est recopié en ADN dans une cellule infectée); enfin, la synthèse de protéine d'après la séquence codante d'un ARN est appelée traduction. Chacun de ces processus implique un ensemble complexe de réactions catalysées par des enzymes (protéines ou ARN), et de multiples interactions ADN–protéine et ARN–protéine. Les protéines, qui constituent le produit final de cet ensemble de processus, sont indispensables à chacune des étapes. Cette chaîne de transfert d'information représente donc le résultat final de l'évolution et non une simple juxtaposition de mécanismes indépendants. Elle se produit dans toutes les cellules vivantes, sans aucune exception. Il n'existe pas d'indication que des cellules vivantes aient pu exister avant que cette chaîne se soit mise en place.

Il est très important de comprendre le concept de "fonction" en biologie, parce qu'il n'a pas d'équivalent en physique. Cette notion semble impliquer une intention, et ceci s'explique bien dans le cadre de l'évolution et de la sélection des organismes vivants. Le fait qu'une protéine donnée, par exemple, n'existe que parce qu'elle remplit une fonction biologique, sans quoi elle n'aurait pas été sélectionnée au cours de l'évolution, est parfaitement étranger à l'expérience des physiciens.

On peut essayer d'expliquer le concept de fonction biologique à un physicien en utilisant l'analogie d'un supraconducteur, un composé chimique qui par sa structure a la propriété très utile ou "fonction" de conduire le courant électrique avec une résistance négligeable. Mais l'analogie est incomplète parce que bien que ce type de composé existe probablement dans la nature, cela ne résulte pas d'une évolution suivie de sélection de matériel qui présente la propriété requise de haute conductivité électrique. (C'est évident si on se rappelle que cette propriété s'exprime uniquement à très basse température!). En fait, pour le développement de matériaux avec des fonctions ou propriétés utiles en physique, l'homme adopte souvent une approche similaire au cycle évolution–sélection. Notamment, dans l'exemple des supraconducteurs, pour trouver des composés à température de transition plus élevée, les chimistes ont adopté une démarche de modifications presque aléatoires des composés pour ne retenir (ou sélectionner) que ceux qui avaient les propriétés les plus intéressantes.

Les méthodes de recherche actuelles en biologie cellulaire sont très éloignées de celles de la physique. C'est cependant avec beaucoup d'intérêt et de plaisir que nous avons pu observer comment l'un des étudiants, un professeur de physique, a pu évoluer dans ses conceptions de la biologie au cours des trois semaines de l'Ecole des Houches. Au commencement du cours, il était convaincu que la biologie était tellement complexe qu'il faudrait des dizaines d'années d'études avant qu'une approche physique dans ce domaine présente quelque espoir de succès. Cependant, au cours des semaines suivantes, il a commencé à entrevoir des mécanismes susceptibles de se prêter à une approche physique, expérimentale et même théorique.

La première partie de ce volume (cours 1 à 4) correspond à une série de cours sur le cytosquelette, ses structures variées et sa dynamique, en particulier au cours du cycle de division cellulaire. Dans ce domaine, l'approche réductionniste est dominante, et les chercheurs s'attachent à analyser la façon dont les structures du cytosquelette se construisent par assemblage des molécules protéiques qui les composent et à comprendre leur croissance et leur écroissance, ainsi que leurs interactions entre elles et avec d'autres éléments de la cellule.

Didier Job décrit les propriétés physico-chimiques fascinantes des microtubules in vitro. Richard Wade s'intéresse à la détermination de la structure de ces microtubules. Richard McIntosh traite le problème du mouvement des chromosomes pendant la division cellulaire. Finalement, Robert Margolis étudie comment le cytosquelette et en particulier les microtubules organisent l'espace dans la cellule.

La seconde partie (cours 5, 6, 7 et résumés dans la section IV) décrit le réseau des membranes internes des cellules eucaryotes et le transport ordonné de protéines qui le parcourt. Ici, l'approche réductionniste se limite jusqu'à présent à la description de facteurs protéiques impliqués dans ce "trafic"; nous sommes encore loin d'une compréhension de ces mécanismes au niveau submoléculaire ou atomique, et il n'est d'ailleurs pas sûr qu'elle nous apprenne grand chose sur les mécanismes généraux. D'un autre côté, les interactions dynamiques membranes–membranes et membranes–vésicules (par exemple fusion entre membranes, et bourgeonnement de vésicules) jouent un rôle essentiel et il serait très utile et très opportun d'améliorer notre compréhension de ces mécanismes par des études physico-chimiques.

Après avoir vu la diapositive d'une cellule pleine d'éléments du cytosquelette pendant la première semaine, l'un des étudiants de l'Ecole était extrêmement surpris de voir une cellule qui semblait entièrement remplie d'un réseau de membranes. Comment pouvait-il y avoir place à la fois pour les fibres rigides du cytosquelette et pour un système dense de replis membranaires? La juxtaposition de ces deux points de vue a illustré pour les physiciens une idée qui paraît évidente au biologistes cellulaires: malgré l'unité fondamentale du monde vivant, que nous évoquions ci-dessus, il existe une grande varriété de types cellulaires, qui diffèrent par leurs stades de développement, leur spécialisation fonctionnelle, etc. Pour analyser un élément cellulaire, on peut donc choisir un type cellulaire où il est particulièrement développé, et appliquer des méthodes appropriées de coloration ou de traitement pour en étudier les divers aspects. Le chapitre principal de cette section est celui de Mary McCaffrey et Bruno Goud. Albert Goldbeter consacre un chapitre au rôle du temps en biologie, et décrit des modèles théoriques permettant d'expliquer les phénomènes d'oscillations biochimiques et cellulaires. De courtes contributions de David Sabatini, Donald Engelman, Benjamin Kaupp et Regis Kelly indiquent le contenu des conférences qu'ils ont données et fournissent des listes de références constituant une bonne documentation.

Comme préambule à son cours, Regis Kelly a insisté sur le fait que l'évolution constitue une différence majeure entre la biologie et la physique, et a indiqué qu'il n'existe que trois approches expérimentales possible en biologie: a) la génétique; b) la reconstitution de systèmes fonctionnels in vitro, et c) la pharmacologie, qui permet d'observer comment des agents spécifiques perturbent un système biologique.

La troisième partie de ce volume traite de l'émergence de niveaux d'organisation supra-cellulaires. Le chapitre de Olivier Pourquié et Frank Bourrat, et celui de Philippe Vernier sont consacrés au développement du système nerveux central. Nous apprenons que l'information génétique, codée dans la séquence de l'ADN ne suffit pas et que l'organisation spatiale d'un œuf joue un rôle essentiel dans le développement d'un organisme multicellulaire (mais à vrai dire, ceci est déjà vrai pour la division d'une simple cellule). Le programme de développement d'un organisme est donc contenu à la fois dans son ADN et dans la géographie interne de l'œuf.

En dehors des cours proprement dits, nous avons eu plusieurs conférences données par les professeurs invités et par les participants. Certains de ces séminaires ont été rédigées et constituent la quatrième partie du volume, avec de brèves indications sur les étudiants qui ont suivi l'Ecole d'Eté.

Le volume se conclut par une réflexion sur la physique et la biologie proposée par Bernard Jacrot. La carrière de recherche de Bernard Jacrot est tout-à-fait exceptionnelle, puisqu'il a apporté des contributions originales importantes à la fois en physique et en biologie, dans les domaines du magnétisme, de la diffraction des neutrons, de la virologie et de la biologie structurale moléculaire. Il a participé activement (à l'organisation et à l'animation) de l'Ecole d'Eté et, dans ce chapitre, il nous fait part de ses idées sur la façon dont physiciens et biologistes abordent les problèmes dans leurs domaines respectifs.

L'Ecole d'Eté des Houches dépend de l'Université Joseph Fourier et de l'Institut Polytechnique National de Grenoble. Elle est subventionnée par le Ministère de l'Education et de la Recherche, le Centre National de la Recherche Scientifique (CNRS) et la Commission à l'Energie Atomique (CEA). La présente session a été financée par l'OTAN, en tant qu'Institut d'Etudes Avancées (Advanced Study Institute, ASI). Nous remercions les membres du Conseil d'Administration des Houches, et particulièrement Jean Zinn-Justin pour leurs conseils (et leur aide) pendant la préparation de ce cours.

Nous remercions aussi Ghislaine Chioso, Isabelle Lelièvre et Brigitte Rousset pour leur participation à la préparation et à l'organisation de la session, ainsi que l'ensemble du personnel de l'Ecole pour leur accueil amical aux Houches.

Giuseppe (Joseph) Zaccai
Jean Massoulié
François David

PREFACE

The July 1996 Summer School at Les Houches was devoted to aspects of Cell Biology for students from other disciplines, especially Physics. But what is Cell Biology? And, while we are asking such questions, what is Physics? Someone said that Physics is what Physicists do. A corollary that immediately leaps to mind is: Physicists do not do Biology. So, some Physicists came to Les Houches to learn about Cell Biology from scientists who do it. Cell Biology is about cells and their organisation: the minimal and universal level of Life.

As well as the cultural pleasure of hearing experts talk about "Life", there were two professional questions behind the Physicists' interest: (i) to find out whether or not they themselves could contribute to Biology and the understanding of Cell function; (ii) to find out if there were interesting Physics problems arising out of the study of Cells. These questions appear to differ mainly by the motivation underlying each one, but they, in fact, cover quite different approaches. This is best illustrated by the example of biological membranes. It was suspected from the 1920's that a lipid bilayer was the main permeability barrier in cellular membranes. Forty years later, systematic physico-chemical studies on extracted lipids revealed the rich diversity of their polymorphism. It is certainly of interest, for the understanding of biological function, to characterise the properties of lipids and their bilayer forming properties. These studies, however, were also precursors of significant progress in the Physical Chemistry of amphiphilic molecules and surfactants, driving aspects of basic and applied research only marginally related with Biology, if at all.

The link between Molecular Biochemistry and Physical Chemistry is firmly established. As a direct result of the "Structure-Function" hypothesis, it is a basic tenet of modern Biology that the function of a biological macromolecule, a protein or nucleic acid for example, is determined by its Physical Chemistry. Biochemical or Structural experiments at the molecular level have a fundamental requirement that defines also the Physical Chemistry boundaries of the problem: the system studied (e.g. a given protein) must be well-defined and circumscribed — a biochemist would say "pure" — and functionally active. A priori it would seem, therefore, that these conditions are absolute prerequisites for the analysis of biologically relevant properties in terms of "molecular" models. As we shall see below, this is not the case in Cell Biology.

Cell Biology is the science of cellular processes. These are usually studied in terms of genetically controlled molecular fluxes and dynamic intracellular structures. A cell is not a complex static system. Active dynamic processes either maintain a certain or guide organised changes such as those that underly embryonic development or adaptation to the environment. The requirement for purity, essential to the biochemical and structural approaches at the molecular level, does not have the same relevance in Cell Biology because all the actors on the cellular stage are far form being known. An important part of many experiments, in fact, is to find out which macromolecules are involved in a given process. If Physics were to come into this picture it would have to be in a very different way than for Structural Biology and Biochemistry at the molecular level.

An important (and essential) contribution of Physics to Biology continues to be in the realm of instrumentation and methods and a large part of progress in Cell Biology is associated with new and powerful Physical approaches such as confocal microscopy, patch-clamp methods or optical tweezers. Instrumentation and methods, however, have been the subjects of various meetings and were not treated in detail at this Les Houches School.

Biologists and physicists have in common that they are comfortable dealing with different levels of organisation and hierarchies. Biologists are interested in spatial dimensions varying from those corresponding to chemical reactions to those of organisms and ultimately relationships between organisms and their environment (ecology) — a range covering more than 10 orders of magnitude (from Å to km) which includes small molecules (amino acids, nucleotides, lipids, sugars, . . .), macromolecules (proteins or nucleic acids), large complexes (ribosomes, multienzyme complexes, . . .), integrated systems (the endoplasmic reticulum, synapses, the Golgi, endosomes, vesicles, . . .) up to cells, multicellular organisation and ecosystems. Similarly, Biology is concerned with a wide range of time scales, from the femtosecond of electronic transitions to the hundreds of millions of years of evolution. Physics of course deals with the full range of time, sizes and organisations, from elementary particles to the universe. It is concerned with fundamental properties best expressed in simple or cooperative systems. At the molecular level of condensed matter, for example, physicists are happiest trying to understand properties of crystals in which an atomic motif is repeated regularly and infinitely, or homopolymers under well-defined thermodynamic conditions. Whereas to a Biologist a crystal is but a device that allows the study to atomic resolution of large and complex molecules such as proteins or nucleic acids, in order to help in the understanding of their function.

Physicists are uncomfortable with the complexity of biological systems. The challenge they face in Biology is how to simplify the description of a system while maintaining its functional relevance. This is difficult even in the reductionist approach of Molecular Biology — each protein being a hetero-polymer with complex properties that are uniquely related to its function. The flow of information in Molecular Biology is from DNA to RNA to protein. This is sketched in the scheme: **DNA** ↔ **DNA** ↔ **RNA** ⇒ **protein**. The process DNA to DNA is replication, DNA to RNA is transcription, RNA back to DNA is reverse transcription (it happens when retroviral RNA is transcribed into the DNA of the host cell), RNA to protein is translation. The fascinating fact is that each of these processes is made up of a complex set of controlled and catalysed reactions involving enzymes (protein or RNA), DNA–protein and RNA–protein interactions. The end product of the scheme (protein) is required for each and every step. The scheme, therefore, represents a stable end-product of evolution and not a set of sequential events. It is valid in all living cells, and there are no exceptions. There is no direct evidence of life forms that existed prior to the scheme. It is very important to understand the concept of "function" in biology because its equivalent does not exist in Physics. The fact that a given protein, for example, exists only because it has a useful biological function (otherwise it would not have been selected by evolution) is perfectly foreign to a physicist's experience. One could try to explain function to a physicist by using the analogy of a superconductor: a certain material has the useful function of conducting electric current with negligible resistance. But the analogy is wrong or at best incomplete because the superconducting material was not selected to "exist" because of its useful properties!

Current approaches in Cell Biology are quite far from Physics. It was interesting and gratifying to observe, however, how one of the students, a Physics Professor, evolved in his view during the three weeks of the School. At the beginning of the course, he was convinced that the complexity was such that decades of biological studies would be required before any hope of usefully applying Physics to the field. As the School progressed, however, he started to perceive patterns that he thought might be amenable to physical experimentation and even theory.

The first section of this volume (courses 1 to 4) corresponds to courses on the cytoskeleton, its various structures and its dynamics, especially during the cell cycle. The reductionist approach is favoured in this field and considerable effort is spent to find out how these structures are built up from their component molecules, how they grow or decrease in size, how they interact with each other and with other cell components. The chapter by Didier Job describes the fascinating in-vitro physical-chemical properties of microtubules. It is followed by a chapter on microtubule structure determination by Richard Wade. Richard McIntosh treats the problem of chromosomes moving apart just before cell division. The section is concluded by Robert Margolis who looks at how the cytoskeleton and microtubules, in particular, create order in the cell.

The second section (courses 5, 6, 7 and course summaries in Section IV) describes the endo-membrane system of a eukaryotic cell and the regulated protein traffic that flows through it. Here, the reductionist approach is so far limited to the description of protein factors; we are as yet far from a submolecular description and, in any case, would it be useful? On the other hand, membrane–membrane and membrane–vesicle interactions (regulated fusion or budding for example) are important and there are ripe opportunities for improving biological understanding from studies of the Physical Chemistry of these systems. One of the students at the School was extremely puzzled after seeing a slide of the interior of a cell that showed it full of cytoskeletal components, in the first week, and another slide in the second week showing a cell absolutely full of membrane structures. How can there be room for both the cytoskeleton and the endomembrane system? The juxtaposition of the cytoskeleton and membrane points of view conveyed important ideas to the physicists' — ideas that are taken for granted by cell biologists: there are different type of cells; they have different stages of development; they can be chosen, stained or treated specifically to display or study different aspects. The main chapter in this section is by Mary McCaffrey and Bruno Goud. Albert Goldbeter provides a chapter on the role of *time* in Biology and theoretical models to describe biochemical and cellular oscillations. Short contributions by David Sabatini, Donald Engelman, Benjamin Kaupp, Regis Kelly describe their lectures and provide reading lists. Regis Kelly started his course insisting on the role of evolution in the difference between Biology and Physics and stating that there are only three experimental approaches in Biology: (1) genetics; (2) in vitro reconstitution; (3) pharmacology to observe how specific reagents affect a system.

The onset of higher levels of organisation is presented in the third section of the Volume. Chapters by Olivier Pourquié & Frank Bourrat and Philippe Vernier, it deals with the development of the central nervous system. Here we learn the limits of genetic information as encoded in the sequence of DNA and the importance of spatial location in an egg that will lead to a differentiated multicellular organism. The programme for an organism is contained in its DNA but also in the internal geography of the egg.

Preface

During the three weeks of the course there were several seminars by lecturers and participants. Some were written up and are included in the fourth section of this Volume, together with short notes on the students who attended the School.

The Volume is concluded by a reflection on Physics and Biology by Bernard Jacrot. Bernard Jacrot has enjoyed research careers in both Physics and Biology and made important, pioneering contributions to the fields of Magnetism, Neutron scattering, Virology and Structural Molecular Biology. He participated in the School and, in this short chapter, shares some of his thoughts about the different ways in which physicists and biologists tackle problems in their respective fields.

Les Houches Summer School is affiliated to the University Joseph Fourier and the National Polytechnic Institute in Grenoble. It is subsidized by the Ministry of Education and Research, the National Center of Scientific Research (CNRS) and the Atomic Energy Commission (CEA). This session has been funded by NATO as an Advanced Study Institute (ASI). We thank the members of the Les Houches Committee and especially Jean Zinn-Justin for their advices during the preparation of this session.

We also want to thank Gislaine Chioso, Isabel Lelievre and Brigitte Rousset for their participation in the preparation and organization of the session, as well as the whole staff of the School for their friendly reception.

Giuseppe (Joseph) Zaccai
Jean Massoulié
François David

CONTENTS

SECTION I. CYTOSKELETON AND CELL CYCLE

Contents

Course 3.   The Cellular Machinery for Chromosome Movement, by J.R. McIntosh   *47*

Course 4.   The Role of Microtubules in the Creation of Order in the Cell, by R.L. Margolis   *71*

SECTION II. INTRACELLULAR COMMUNICATION
Membranes – Synapses – Time

SECTION III. DEVELOPMENT OF THE CENTRAL NERVOUS SYSTEM

Contents

SECTION IV. Lectures and Seminars Presented at the Summer School but not Published in the Proceedings

Student Seminar 2. Isolated Nerve Cell Response to Laser
Irradiation and Photodynamic Effect,
by A.B. Uzdensky   *227*

Short Reports on Ph.D. Student Seminars   *243*

SECTION V. CONCLUSION

Relations between Physics and Biology,
by B. Jacrot   *253*

SECTION I. CYTOSKELETON AND CELL CYCLE

COURSE 1

MICROTUBULE DYNAMICS IN VITRO AND THEIR RELATIONSHIP WITH CELLULAR FUNCTION

Didier Job

*Département de Biologie Moléculaire et Structurale, Laboratoire du Cytosquelette,
Institut National de la Santé et de la Recherche Médicale Unité no. 366, Commissariat à
l'Energie Atomique de Grenoble, 17 rue des Martyrs, 38054 Grenoble Cedex 9, France*

G. Zaccai, J. Massoulié and F. David, eds.
Les Houches, Session LXV, 1996
De la Cellule au Cerveau
From Cell to Brain: Intra- and Inter-Cellular Communication –
The Central Nervous System

Contents

1. Introduction

Living organisms are comprised of prokaryotes and eukaryotes. Prokaryotes such as bacteria are devoid of nuclei. The genetic information is encoded in a DNA molecule, free in the cytoplasm. In eukaryotes, on the other hand, the DNA molecules are packaged in the cell nucleus.

The genetic material in eukaryotic cells is much more abundant and complex than in prokaryotic cells. However, eukaryotic cells and prokaryotic cells have an essential function in common: their survival depends critically on the even segregation of the genetic material in daughter cells, following cell division (called mitosis). In bacteria, the correct segregation of the newly replicated and of the parental DNA molecules among daughter cells depends on the attachment of both DNA strands to specific sites on the bacterial membrane. These attachment sites are placed on either sides of the cell wall which eventually separates the two daughter cells. Eukaryotic cells with their numerous chromosomes and their enormous amount of genetic material had to develop a more sophisticated mitotic machinery to achieve correct segregation of DNA during cell division. Microtubules are essential components of the elaborate mitotic machinery of eukaryotic cells. Microtubules are specific of eukaryotic cells as compared to prokaryotic cells. The appearance of microtubules during evolution is a landmark of the transition between prokaryotes and eukaryotes as is the appearance of the cell nucleus itself [2,3].

Mitosis in eukaryotes depends on two main functions of microtubules: the apparent ability of microtubules to form highly organized structures and the capacity of microtubules to generate organelle movements: these two major microtubule functions, organizing the cell interior and generating motility, have been exploited by cells in an amazing variety of ways [4].

The precise ways through which microtubules achieve cell organization and motility are not fully understood yet. A central hypothesis in the present article is that microtubule functions depend fundamentally on microtubule dynamic properties. In the first section of this paper, the intrinsic dynamics of pure microtubule preparation *in vitro* are briefly described. Subsequent sections deal with the possible relationships between microtubule dynamics as observed *in vitro* and microtubule functions.

7

2. Basic microtubule dynamics (for review, see [4])

2.1. Tubulin and microtubules

Microtubules are cylindrical aggregates of a single heterodimeric protein called tubulin. Tubulin addition to and loss from existing microtubules occur only at microtubule ends. The aggregation of pure tubulin to form microtubules can be induced in a variety of conditions but two factors play an important role.

First, pure tubulin microtubules do not form at low temperature. Therefore, tubulin aggregation is almost always induced by warming up tubulin solutions to 30°C–37°C.

Secondly, tubulin is incorporated into microtubules as a complex with GTP. GTP belongs to a class of molecules that comprise a triphosphate group. Hydrolysis of the third phosphate (to form GDP + Pi) yields free energy which can be used to catalyze coupled biological reactions. GTP is essential to trigger tubulin polymerization. Following warming and GTP addition to soluble tubulin dimers, tubulin polymerization proceeds in three phases, as measured by turbidimetric methods.

The first phase is silent and is usually referred to as the nucleation phase. During this phase, ill defined microtubule "suds" are formed.

The second phase is that of polymerization. During this phase, nascent microtubules elongate through tubulin-GTP addition onto their ends. Tubulin-GTP interaction with microtubule ends triggers the appearance of tubulin GTPase activity. GTP is hydrolyzed to form GDP. Therefore, microtubules are formed of tubulin-GDP. The GDP molecule is a structural component of the microtubule wall, GDP cannot be exchanged with external free GTP or GDP once incorporated into microtubules.

The third phase is that of steady state. At steady state, a fixed proportion of tubulin is in the polymeric form while another tubulin pool remains soluble.

The concentration of free tubulin dimers at steady state varies with buffer conditions and corresponds to the so-called tubulin critical concentration for microtubule assembly. Microtubule assembly has two main features: it is a reversible phenomenon and, at steady state, the tubulin microtubule system remains out of equilibrium.

Microtubule assembly is reversible: a temperature drop below 10°C–15°C, removal of GTP, dilution of microtubule suspensions and subsequent decrease of the free tubulin concentration, all result in rapid microtubule depolymerization to generate free tubulin-GDP molecules. Microtubule disassembly can also be triggered by adding specific drugs to a microtubule solution, such as colchicine, vinblastine or nocodazole. These drugs poison tubulin assembly onto microtubule ends and thereby induce microtubule depolymerization.

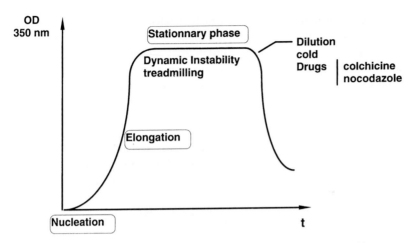

Fig. 1. Schematic representation of the various phases of microtubule assembly.

2.2. Steady state microtubules

At steady state, microtubules remain out of equilibrium. Continuous GTP hydrolysis and thereby energy consumption, is required for the maintenance of steady state microtubules, which apparently use energy to generate active exchanges between their constitutive subunits and the tubulin molecules of the soluble pool. Two mechanisms are thought to account for such tubulin exchanges.

The first has been called "treadmilling". Treadmilling relies on an extraordinary property of microtubules: at steady state, microtubule ends behave differently [12]. One end, (the +end) continuously incorporates new tubulin dimers while the other end (the −end) continuously looses them. Such asymmetric behaviour clearly requires energy consumption. However, the precise mechanisms which couple GTP hydrolysis to the generation of the functional asymmetry in microtubules are unknown, so far. Microtubules are structurally asymmetric, polarized, polymers (see the course by Wade). Functional and structural asymmetries of microtubules may represent, in principle, completely independent properties and might be unrelated. However,it appears that both asymmetries are coupled. In cells, all microtubules have the same polarity. Their −ends are central, embedded in a microtubule organizing center, while their +ends are peripheral, free in the cell cytoplasm [14]. When microtubules are free in solution, treadmilling results in an apparent polymer migration actually resulting from the propagation of a wave of tubulin polymerization–depolymerization. When microtubules are stationary, treadmilling generates an apparent subunit flux through the polymers of tubulin molecules travelling from the +ends toward the −ends [13].

The second mechanism through which steady state microtubules exchange their constitutive tubulin with soluble tubulin relies on another strange microtubule property known as "dynamic instability". Steady state microtubules show spontaneous length fluctuations due to successive phases of disassembly and reassembly [7,15]. The precise processes underlying dynamic instability are unclear. All current models rely on the idea that microtubules ends are structurally different during assembly and disassembly, with a difficult transition between the assembly and the disassembly conformations.

Reversible assembly, treadmilling, and dynamic instability summarize the basic properties of microtubules assembled from pure tubulin *in vitro*. In living cells, microtubules generate movements and organization. The issue addressed in the following sections concerns the potential relationships between the basic physico-chemical properties cellular functions of microtubules.

3. Microtubules and the generation of movement: the polymer biased diffusion model

3.1. Anaphase chromosome movement (see also the course by McIntosh)

One of the basic "raisons d'être" of microtubules is to participate in mitosis. Anaphase is the phase of mitosis during which chromosomes are segregated in the dividing cell, through migration from an equatorial position toward the two opposite cell poles. During anaphase, microtubules form two hemi-spindles, connecting the cell poles to the migrating chromosomes. A crucial observation is that poleward chromosome migration is accompanied by concomitant shortening of spindle microtubules [4]. A natural question which arises concerns the existence of plausible physico-chemical mechanisms which could couple microtubule disassembly and chromosome movement. Such coupling can, in principle, be achieved through a mechanism known as "polymer guided" [5], or "polymer biased" [8] diffusion. Polymer biased diffusion relies on a cycle of microtubule disassembly and fluctuation followed by chromosome diffusion. For the sake of simplicity, we shall consider the case of one microtubule connected to one chromosome. Initially, the microtubule is at rest, with one end connected to the cell pole and the other attached to the chromosome. The exact geometry of the attachment sites is not known. Here, we shall suppose that microtubules are attached to cell structures through interaction of a terminal tubulin subunit with high affinity binding sites at the cell poles and/or on the chromosome. Other types of connections are obviously plausible but uncertainties in this matter are of no consequence with regard to the general principles of biased diffusion. Let us suppose that the terminal tubulin molecule connecting the microtubule to the cell pole dissociates from

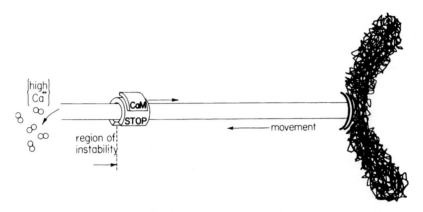

Fig. 2. Polymer guided diffusion: depiction of a hypothetical anaphase structure. Microtubules are attached to the cell pole through a ring structure formed of STOP proteins. STOP is a calmodulin regulated protein which can both stabilize microtubules and slide along the microtubule wall. Its role is to guide the polymer displacement and to hold the polymer laterally. STOP also stabilizes the polymer region lying between it and the kinetochore so that the microtubule remains polymerized in a metastable manner until beyond the STOP, in a region where its instability increases rapidly due to high local calcium concentrations. Depolymerization is rapid beyond the STOP and, consequently, diffusion of the polymer and the kinetochore is inexorably unidirectional and poleward.

the polymer. Because of Brownian motion, the polymer undergoes rapid length fluctuations. Some of these fluctuations will bring a terminal tubulin subunit of the polymer in contact with the high affinity binding site at the cell pole, resulting in binding. The polymer is now frozen in a slightly extended state. The chromosome, on the other hand, is also moving spontaneously through diffusion. If not attached to a microtubule, the chromosome would move randomly, with very little net displacement. But the energy involved in the diffusion of the chromosome is lower than the energy corresponding to the return of the polymer to its equilibrium length. Therefore, instead of diffusing at random, the chromosome will diffuse toward the pole bringing the system to a new equilibrium state. Then, a new cycle of depolymerization–fluctuation–diffusion can begin and successive cycles will finally drive the chromosome to the pole.

There are, of course, kinetic constraints involved in the basic cycle (for a complete discussion see [5]). For instance, depolymerization should be slow enough to leave enough time for chromosome diffusion to proceed. One way to illustrate such constraints is to consider the system following conversion of nanoseconds into seconds and of micrometers into meters. Using these units and an adequate set of kinetic parameters, the chromosome would be initially 10 m away from the cell pole. The microtubule attaching the chromosome to the pole would loose one subunit each 30 yr. Rebinding of the microtubule end to the pole, following tubulin dissociation, would occur in a second. The chromosome would diffuse over 1 cm in 3 yr. In such conditions, the chromosome migration to the pole would be completed within a period of 30000 yr. This looks like very slow motion. It is somehow remarkable that such slow motion is actually much faster than real chromosome motion during anaphase [5].

3.2. Biased diffusion in the real world

Whether or not anaphase can result from biased diffusion in cells, in the absence of any superimposed energy dependent source of motion is still a matter of controversy. However, poleward chromosome movement by biased diffusion has been demonstrated in reconstituted systems, *in vitro* [8,9]. Furthermore, polymer biased diffusion is a general "push–pull" mechanism which applies to different systems. For instance, it is now generally accepted that the bacterium *Listeria* moves inside cells through polymer guided diffusion. Following entry into cells, *Listeria* recruits actin to form a flagellum. Actin polymerization occurs at the junction between the flagellum and the bacterial wall. The bacterium body migrates within the cell at a rate matching that of actin polymerization [23].

4. Microtubules and self organization

Microtubules form organized arrays in cells. It seems hardly possible that the spatial arrangement of microtubules is specified by some sort of central organizer in the cell. A more likely possibility is that microtubules have self organizing capacity. Soon after the discovery of microtubules and the realization that microtubules were dynamic polymers, it was intuitively felt by researchers in the field that microtubule organization was a consequence of microtubule dynamics, modulated by cell regulators. The question addressed in the next sections of this paper concerns the consistence of such a view with basic physico-chemical principles and the possibility to assay microtubule self organization, *in vitro*. Tools and concepts are different whether one considers nascent, rapidly growing, microtubules or steady state polymers. Rapidly growing microtubules, as observed during the establishment of the microtubule network, obey simple linear kinetics. Steady state microtubule kinetics are clearly more complex.

4.1. Self organization of rapidly growing microtubules: a diffusion based model [19]

During the establishment of microtubule networks, polymers grow rapidly from a single cellular organelle, the centrosome. It is a common observation that micro-

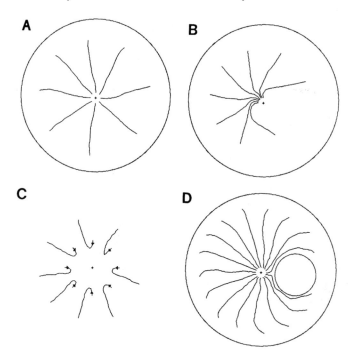

Fig. 3. Self organization of the microtubule network according to a diffusion based model. Results of a computer simulation. (A) The 8 starting positions of the microtubules were equally distributed on a circle of radius 0.3. The initial direction of growth is radial, $dt = 0.04$. Ten iterations (computed elementary spatial steps) have been performed. (B) Same as (A), except that the eight starting positions were equally distributed on a quarter of the circle. (C) The 8 starting positions were equally distributed on a circle of radius 0.3. The initial direction of growth is inward, toward the circle's center. Fifteen iterations have been performed. (D) Sixteen starting positions were equally distributed on a circle of radius 0.1. The initially radial direction of growth was randomly perturbed. Twenty iterations were performed. The flux of tubulin through the boundary of the obstacle is supposed to be nil.

tubules growing from centrosomes form regular arrays, made up of approximately linear polymers, evenly distributed in the cell interior. What are the driving forces underlying such regular microtubule organization? Intrinsic microtubule rigidity may look as an important determinant but it is known that, in some situations, cellular microtubules can have a highly tortuous configuration. In the absence of other strong organizing forces, the vector of microtubule growth can only be governed by diffusion laws. Such laws are well established.

The behaviour of polymers growing in a finite medium and originating from a small region has been modeled on this basis. At time zero, the tubulin concentration is assumed to be homogeneous. The rate of polymer growth is constrained to be proportional to the local tubulin concentration, in the vicinity of the micro-tubule growing end. The only dynamic processes involved in the model are microtubule growth and tubulin diffusion. The equations which describe microtubule behaviour in such conditions depend on a small number of parameters whose values have been experimentally determined in cells. Computer simulations showed that growing microtubules simply obeying diffusion laws form nicely organized networks. Polymers grow linearly, are regularly spaced and skirt obstacles. Therefore, elementary laws of diffusion suffice to confer self organizing properties on rapidly growing microtubules.

The organizing force in such a diffusion based model, resides in the tubulin concentration gradients created by rapid microtubule growth. Once established, microtubule networks look quasi stationary, hence organizing tubulin gradients would be expected to vanish and the whole network should degenerate into chaotic arrays of tortuous polymers. Cells actually contain a small number of stable, dynamically inactive microtubules. Such inert microtubules are conspicuously tortuous and knotty. These observations raise questions with regard to the capacity of steady state microtubules to generate tubulin concentration gradients in order to maintain organization. These questions are addressed in the next sections.

4.2. Microtubule oscillations

In principle, maintenance of quasi stationary chemical gradients and self organization can occur in dissipative, non linear chemical systems. Dissipative systems continuously consume energy, are thermodynamically irreversible, and are kept out of equilibrium. With this definition, microtubule solutions clearly represent dissipative systems. Microtubules continuously incorporate tubulin-GTP molecules, which results in energy consumption through GTP hydrolysis. The reaction is thermodynamically irreversible since tubulin-GDP dissociation from microtubules does not generate tubulin-GTP. Finally, as discussed above, microtubules are out of equilibrium at steady state due to dynamic instability and treadmilling. Do microtubules behave as a non linear system? We do not have, as yet, precise knowledge of the equations that describe microtubule behaviour. But, non linearity of microtubule behaviour has been demonstrated experimentally. A salient feature of non linear systems is their capacity to undergo oscillatory behaviour, when placed in the proper conditions. Depending upon the conditions under which polymerization is initiated, tubulin can assemble either following monotonic kinetics or adopt an oscillatory mode [17]. When present, microtubule oscillations involve successive and spontaneous phases of extensive (sometimes complete) microtubule disassembly followed by polymer reassembly, until, progressively, a steady state is reached. Microtubule oscillations can be triggered in different ways. A basic requirement for such oscillations to occur is that the rate

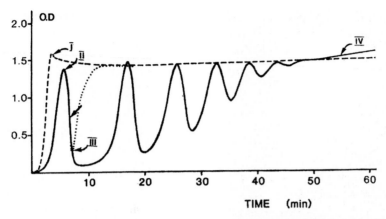

Fig. 4. Microtubule oscillations. Dashed curve: microtubule assembly in the presence of 1 mM free GTP. In the presence of excess GTP, tubulin assembly is monotonic. Full curve: microtubule assembly in the absence of free GTP. In this case, tubulin-GTP was generated from tubulin-GDP under the action of a regenerating system. In this condition, the rate of generation of tubulin-GTP becomes a limiting factor for tubulin assembly and this induces microtubule oscillations (see text).

of initial tubulin assembly should be rapid as compared to the rate of tubulin GTP regeneration from tubulin GDP. This can be achieved either by controlling the rate of GTP regeneration from GDP [17], or by assembling tubulin at very high rates, in optimal buffer conditions and at high tubulin concentrations [1]. The exact molecular mechanisms underlying microtubule oscillations is still debated. But, in any case, the existence of microtubule oscillations is solid evidence that some of the chemical reactions involved in microtubule formation are non linear.

Clearly, microtubule assembly results from dissipative and non linear chemical reactions. Such systems can, in principle, generate patterns known as dissipative space structures or Turing structures. The next section of this article deals with experimental evidence that microtubule solutions can indeed self-organize into such structures.

4.3. Pattern formation in microtubular solutions [20,11,6]

Pattern formation in microtubular solutions are easy to observe. Microtubule suspensions are turbid. Tubulin solutions are limpid prior to assembly and turn white during microtubule formation. It is a common observation that microtubule suspensions are not always homogeneously turbid. Instead, they often show distinct clear and turbid domains. Until about a decade ago, the significance of such spontaneous pattern formation has been overlooked. Since, however, pattern formation in microtubular suspensions has been subjected to systematic investigation. Initial

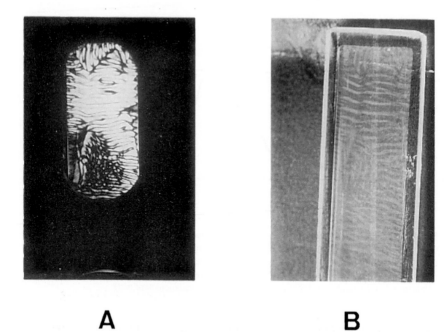

A   **B**

Fig. 5. Microtubule dissipative space structures. Pure tubulin was assembled in a spectrophotometer cell (40 × 10 × 1 mm). (A) Observed in reflected light, the microtubule solution showed a pattern formed of successive dark and white stripes. (B) Same microtubule solution observed through crossed polarizing filters.

observations concerned pure tubulin assembled into microtubules in spectrophotometric cells. Following microtubule formation, macroscopic striped patterns arose [20,6] consisting of a series of horizontal white and dark stripes encompassing the whole cell. Each stripe was about 2 mm wide. The spatial arrangement of microtubules within the stripes was determined using small angle neutron scattering which made available the appropriate spatial resolution (∼10 Å). The stripe pattern was shown to result from periodic changes in microtubule orientation throughout the cell. In each stripe microtubules are aligned, but microtubule orientation shifts from −45° to 135° to the horizontal in successive stripes [20]. Pattern formation was not observed in stable, non dynamic, microtubule solutions, and was modified when the energy dissipation of the system was modified by addition of variable concentrations of Mg^{2+}. Therefore, it was proposed that microtubules were forming dissipative space structures [20]. Self-organization is a salient feature of living organisms [16] and it has been supposed for years that biological systems should have the capacity to form Turing structures [18]. Microtubules appeared to provide a biochemically simple system endowed with such capacity. During the past few years, the possibility that microtubules were forming true Turing structures has been tested in different ways. Pattern formation through reaction-diffusion is, in principle, sensitive to weak fields and can show bifurcations when perturbed at critical time points. Pattern formation in microtubular solutions has been shown to be sensitive to gravity [20]. When tubulin is polymerized in vertical spectrophotometric cells, a stripe pattern arises. But when tubulin is polymerized in horizontal cells, no stripes are formed. Centrifugation of horizontal cells at low speed results in restoration of the stripe pattern. These results strongly suggested sensitivity of microtubule organization to gravity. This possibility is now under direct test in true low gravity conditions (J. Tabony, personal communication). The sensitivity of microtubule spatial organization to gravity was further used to show bifurcation properties in microtubular solutions. When tubulin is initially placed in vertical cells subsequently shifted to horizontal position prior to occurrence of any striped pattern, a normal striped pattern arises hours later. Similarly, no stripes form in samples initially placed in the horizontal position and shifted to the vertical, at early stages of tubulin polymerization. These data suggest the existence of bifurcation points early during the process of microtubule pattern formation (Tabony). Such bifurcation properties confer memory to the system. Finally, very recently, the existence of tubulin concentration gradients in self-organized microtubular solutions has been directly demonstrated ([22] and J. Tabony, personal communication). The occurrence of such gradients is a clear indication of reaction-diffusion processes. Therefore, their occurrence in microtubular solutions is very strong support that microtubules form true Turing structures.

5. Conclusion

Moving organelles and generating order seem to be the main functions of microtubules in cells. It is somewhat remarkable that intrinsic microtubule dynamic properties seem to suffice to account for microtubule functions. Schematically, reversible assembly is probably sufficient for generation of organelle movement through biased diffusion while complex superimposed steady state microtubule dynamics confer self-organizing properties to microtubular solutions. However, to what extent and how cells make use of intrinsic microtubule properties is not clear, as yet. Globally, cytoplasmic microtubules behave as pure tubulin microtubules assembled *in vitro*. Cytoplasmic microtubules show reversible assembly capacity, active subunit turnover, rapid treadmilling, and length fluctuations. But a major difference between cytoplasmic microtubules and pure tubulin microtubules is that, in the absence of cell regulations mechanism, cytoplasmic microtubules are dynamically inactive, due to association with stabilizing proteins. Therefore, normal microtubule dynamics in cells results from the action of cellular enzymes acting on a template of stable polymers. It is a common situation in biology that the basic physico-chemical properties of a system are both indispensable to function and masked by powerful superimposed cell regulations. It is likely that basic microtubule dynamics are both tightly controlled and optimized in cells. A great deal of work remains to be done to begin understand how cells chose and sense a particular mode of cytoskeletal organization.

References

[1] M.F. Carlier, R. Melki, D. Pantaloni, T.L. Hill and Y. Chen, Proc. Natl. Acad. Sci. **84** (1987) 5257.
[2] T. Cavalier-Smith, Ann. N.Y. Acad. Sci. **503** (1987) 17.
[3] T. Cavalier-Smith, Nature **256** (1975) 463.
[4] P. Dustin, *Microtubules*, 2nd Ed. (Springer Verlag, 1984).
[5] J.R. Garel, D. Job and R.L. Margolis, Proc. Natl. Acad. Sci. **84** (1987) 3599.
[6] A.L. Hitt, A.R. Cross and R.C. Williams, Jr., J. Biol. Chem. **265** (1990) 1639.
[7] T. Horio and H. Hotani, Nature **321** (1986) 605.
[8] D.E. Koshland, T.J. Mitchison and M.W. Kirschner, Nature **331** (1988) 499.
[9] V.A. Lombillo, R.J. Stewart and J.R. McIntosh, Nature **373** (1995) 161.
[10] E.M. Mandelkow, G. Lange, A. Jagla, U. Spann and E. Mandelkow, EMBO J. **7** (1988) 357.
[11] E. Mandelkow, E.M. Mandelkow, E.M. Hotani, B. Hess and S. Muller, Science **246** (1989) 1291.
[12] R.L. Margolis and L. Wilson, Cell **13** (1978) 1.
[13] R.L. Margolis and L. Wilson, Nature **293** (1981) 705.
[14] J.R. McIntosh, U. Euteneuer and B. Neighbors, *Intrinsic Polarity in Microtubule Function* (Elsevier Biomedical, 1980).
[15] T. Mitchinson and M. Kirschner, Nature **312** (1984) 232.
[16] G. Nicolis and I. Prigogine, *Self Organization Non Equilibrium Systems* (Wiley, 1977).
[17] F. Pirollet, D. Job, R.L. Margolis and J.R. Garel, EMBO J. **6** (1987) 3247.
[18] I. Prigogine and I. Stengers, *Order out of Chaos* (Heinemann, 1984).
[19] C. Robert, M. Bouchiba, R. Robert, R.L. Margolis and D. Job, Biol. Cell **68** (1990) 177.
[20] J. Tabony and D. Job, Nature **346** (1990) 448.
[21] J. Tabony and D. Job, Proc. Natl. Acad. Sci. **89** (1992) 6948.
[22] J. Tabony, Science **264** (1994) 245.
[23] J.A. Theriot, T.J. Mitchison, L.G. Tilney and D.A. Portnoy, Nature **357** (1992) 257–260.

COURSE 2

STRUCTURE AND FUNCTION OF TWO MOLECULAR MOTORS AND THEIR PATHWAYS

R.H. Wade

Institut de Biologie Structurale (CEA-CNRS), 41 Avenue des Martyrs,
38027, Grenoble Cedex 1, France

G. Zaccai, J. Massoulié and F. David, eds.
Les Houches, Session LXV, 1996
De la Cellule au Cerveau
From Cell to Brain: Intra- and Inter-Cellular Communication –
The Central Nervous System

Contents

1. Introduction

The basic building blocks of complex eukaryotic organisms are cells. These contain a fluid medium, the cytoplasm, surrounded by a lipid bilayer known as the plasma membrane. The cytoplasm is packed with membranes and proteins in the form of organelles, lipid vesicles and soluble proteins interlaced by a web of filamentary proteins called the cytoskeleton. The cytoskeleton has three main components, thin filaments, intermediate filaments and microtubules and these interact to form a scaffold that is involved in structuring the cytoplasm. Thin filaments and microtubules are both used as pathways by specific associated mechanochemical enzymes known globally as motor proteins. This short review concerns F-actin/myosin and microtubules/kinesin, two pathway/motor systems playing many essential life sustaining roles. There is a considerable amount of intensively active ongoing research relating to the biochemistry, the physical chemistry, the biophysics, the structure and the molecular mechanisms involved in the function of the motor proteins, consequently it is only possible to give a brief survey of the properties and structure of the proteins involved in the two pathway/motor systems. Similarly there is a huge literature that can only be inadequately dealt with here, many, but not all, of the cited references are reviews or comments from the past few years. References to original work, to older and perhaps classical articles will be found in these.

The study of skeletal muscle has a long history and models of muscle as overlapping filaments of actin and myosin were already established in the early 1950s. At that time the study of microtubules and associated proteins was still in its infancy as witnessed by the following quotation from J.D. Watson in the 1965 edition of The Molecular Biology of the Gene: *The regular lining up of chromosomes during the metaphase stage is accompanied by the appearance of the spindle. This is a cellular region, shaped like a spindle, through which the chromosomes of higher organisms move apart during the anaphase stage. Much of the spindle region is filled with long, thin, protein molecules, which some people think are similar to the contractile proteins of muscle. If this resemblance is genuine, then perhaps the same mechanism that underlies the contraction of muscles also underlies the movement of chromosomes through the spindle.*

2. General properties of muscle and its proteins

Muscle fibres make up about 40% of the total weight of the vertebrate body. Skeletal muscles are only designed to pull so that two muscles working in opposition are needed to provide complete movement. Skeletal muscle has been studied at length to try to understand the molecular events at the basis of movement. Muscle fibres examined in the light microscope have a regular striated appearance. In the 1950s thin section electron microscopy showed that myofibrils from muscle are made up of end-to-end alignments of basic units called sarcomers within which two distinctly different sets of filaments, the one thick and the other thin, could be seen to interdigitate in a regular fashion. Heads project from the thick filaments and attach to the thin filaments. The thick and thin filaments are made up principally of myosin and actin respectively. Each extended sarcomer is some 2 to 3 microns long and can contract roughly twofold. Later, electron micrographs of longitudinal sections of insect flight muscle, chosen because of their exceptional regularity, showed that the angle of the crossbridges changed by 45° between the rigor and the resting states. Based on this, a sliding filament model of muscle contraction was proposed in which force is generated by the rotation of myosin crossbridges around their attachment site to thin filaments, see [93]. A steric blocking model appeared later to complete this picture. Ca^{2+} released into the sarcoplasm binds to troponin C on the actin filament and mediates a shift of tropomyosin in the groove of the actin filament to uncover myosin binding sites. To go beyond this highly simplified presentation a useful recent review is by Squire [88]. At the present time active research continues on many aspects of the actin–myosin system [11,44,77].

2.1. Actin

Thin filaments are ~9 nm thick polymers of actin with other associated proteins. The actin monomer (G-actin) was first purified from muscle in the 1940s. Starting in the late 1960s, actin began to be detected in non-muscle cells such as blood platelets and brain cells. It turns out to be one of the most abundant proteins in eukaryotes, making up 5–10% of the total protein in many cells rising to a high of around 20% in muscle. G-actin is a 375 amino-acid globular protein with a molecular weight of ~42 kDa that binds adenosine triphosphate (ATP) and divalent cations such as Mg^{2+} and Ca^{2+}. It is highly conserved throughout species and is most probably derived from a single ancestral gene. There are several actin isoforms. The α-actins are specific to skeletal and smooth muscle each of which has two actins that are identical to within 4–6 amino-acids. There are also the non-muscle actins. The β and γ isoforms may differ by up to 25 amino-acids. As an example of conservation across species the actins in slime mould and rabbit muscle are identical to within 17 amino-acids. Mutations of actin can be responsible for genetically related muscular defects.

G-actin can be obtained in a highly purified form by repeated steps of polymerisation and depolymerisation. It is stable at low ionic strengths whilst at physiological ionic strengths it self-assembles into filamentary F-actin, similar to thin filaments. Assembly in the test tube requires ATP, Mg^{2+} ions, 0.1 M KCl. The growth of actin filaments is found to be ten times faster at one end, the plus end, than at the other. Although ATP hydrolysis is not required for assembly, the ATP associated with each actin monomer is hydrolysed during the assembly process, so that actin filaments are mostly made up of actin-ADP. The critical concentration required for assembly is about 0.2 μM and this is considerably smaller than the concentration of G-actin in most cells. X-ray fibre diffraction and electron microscopy shows that F-actin is a 9 nm diameter helical polymer of G-actin with a rise of 2.75 nm and a rotation of almost 180° between monomers, Fig. 1(a).

A family of actin related proteins (arps) can be defined from sequence similarities [25]. At least twenty such proteins have been identified although information on function is still fragmentary. The best known, arp1, associates with microtubules as a component of the dynactin complex. We will see later that the three dimensional structure of actin is practically identical to those the unrelated heat shock proteins and hexokinase despite the absence of sequence similarities. Thus it appears to also belong to a group of proteins with similar tertiary structures but little or no sequence similarity.

2.2. Myosin

Myosin was discovered in skeletal muscle along with actin and was isolated as an F-actin stimulated ATPase. It forms bipolar filaments in muscle and in the contractile ring during cell division. It is a complex protein made up of six polypeptide chains, with an overall molecular weight of some 500 kDa. It has a pair of heavy chains and two pairs of light chains. The dimer formed by the heavy chains has a distinct structure with two globular heads joined by a long α-helical coiled-coil region. Myosin is found at high concentrations in muscle cells (myosin II) and in smaller amounts in other cells. In the test tube and at physiological ionic strengths it spontaneously assembles into bipolar filaments, similar in appearance to the thick filaments of the sarcomer. It is an ATPase that binds to F-actin with a 200-fold increase, up to 5–10 hydrolysis events per second, in ATPase activity compared to the unbound state. E.W. Taylor showed that actin has a high affinity for myosin and for myosin-ADP-P_i and a lower affinity for myosin-ATP. This leads to the idea that during the ATP hydrolysis cycle, myosin can alternatively bind and release from F-actin. In this way force can be generated in the sarcomer through many individual actin/myosin cross bridges giving directed movement be-

a) F–actin

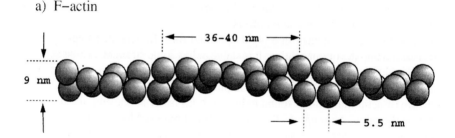

b) microtubule

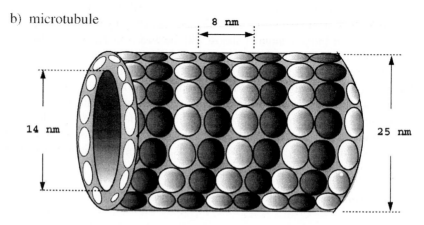

Fig. 1. (a) Filamentary actin (F-actin) is an approximately 9 nm wide polymer in which the globular actin (G-actin, represented by spheres) self-associates to form a helical arrangement. The filament can be considered as two intertwined protofilaments forming a right-handed long pitch helix. In muscle, tropomyosin is located in the groove between the two protofilaments. (b) Microtubules are about 25 nm in diameter and can be considered as cylinders with walls built from a staggered lengthwise array of protofilaments. Each protofilament is a head-to-tail assembly of the tubulin heterodimer, shown as dark and lighter spheres. The commonly accepted surface lattice, B-lattice, is shown here. The number of protofilaments is usually 13 in vivo although there are many exceptions. Microtubules do not necessarily have full helical symmetry.

tween the thick and thin filaments capable of producing the considerable length changes observed in skeletal muscle.

Many new members of the myosin family have been discovered recently by sequence comparisons with the head domain of myosin II (conventional myosin) [8,15,22,29,34,61,63]. More than ten of these so-called unconventional myosins have been identified over the past few years. Several are smaller than myosin II and have a single head. This is the case of myosin I, the brush border myosin found in microvilli. There is evidence that class V myosins are involved in organelle movement [94] and that myosin I might play a role in endo- and exo-cytosis. Myosins are important in sensory systems like hearing, balance and vision, for example myosin III (ninaC) has a role in the electrophysiological response to light. At present much remains to be discovered about these myosins, but they do appear to have important cellular functions and as mentioned below some have been identified as being involved with genetic diseases. The tail domains are probably related to specific functionality and differ between members of the myosin superfamily.

The light chains belong to the family of EF hand proteins, like calmodulin and troponin-C [95,96]. They regulate the ATPase activity of myosin in a Ca^{2+} dependent way. This poses a problem as to how conformational changes in the neck region can be transmitted to the motor domain active site that is about 100 Å away. There is also an idea that the light chains bind to the heavy chain to rigidify the long α-helical region connecting the S1 fragment and the coiled-coil. This could act as a lever arm amplifying small conformational changes near the nucleotide binding cleft.

Mutations in myosin have been shown to be associated with genetic illnesses. Mutations in the β myosin II heavy chain are associated with familial hypertropic cardiomyopathy (FHC) [101]. A single mutation at position 403 (Arg403Gly) disrupts the sarcomer and is responsible for a severe clinical phenotype that occurs in one in five hundred individuals and is quoted as the leading cause of sudden cardiac death in young athletes. Mutations in three other proteins, cardiac troponin T, α-tropomyosin or cardiac myosin binding protein are also associated with FHC leading to the idea that this is a sarcomer specific disease, because, to date, all proteins involved are found in the sarcomer. For example, out of forty reported mutations of the myosin heavy chain all but one are located in the head or near the head to rod junction. Other myosins, myosin VI and VII, have been identified as being involved in genetic deafness and it has been shown that one of these, myosin VI, is concentrated at the base of the stereo-cilia in the hair cell of the inner ear [7,27,36,37].

3. Microtubule structure and organisation

3.1. Tubulin

Microtubules are essential components of eukaryotic cells in which they are involved in the organisation of the cytoplasm and in cell division. Microtubules are ~25 nm diameter structures built from the ~100 kDa tubulin heterodimer,

Fig. 1(b). Tubulin is usually purified from the brain where it makes up about 20% of the soluble protein. Tubulin assembly is temperature sensitive and cycling between 37°C and 4°C is an important step in purifying this protein. There are five or six isoforms of both the α and β subunits, the main differences between these is found in the highly acidic C-terminal region. Both subunits have binding sites for guanine triphosphate (GTP) but only the E-site on β-tubulin allows hydrolysis and exchange of GDP by GTP. As with actin, hydrolysis of the nucleotide triphosphate is intimately connected to assembly of the polymer but is not in itself necessary for assembly. Microtubules are polar structures as witnessed by the different growth rates of the two ends when microtubules self-nucleate and assemble in the test-tube in the presence of GTP and Mg^{2+} and at a temperature above $\sim$30°C. The so-called plus end grows about 10 times more quickly than the minus end. In the cell microtubules are usually nucleated in the region near the centrosome, the microtubule organising centre (MTOC), they grow outwards with the plus end leading and their minus ends usually remain attached to the MTOC.

3.2. Microtubule structure

The structure of microtubules is a good deal more complex than F-actin. It is important to realise at the outset that the α and β subunits of the tubulin heterodimer have very similar amino acid sequences and three dimensional structures [69]. As a result, some properties of microtubules can be understood at the monomer level whilst in other cases subtle differences in the structure of the subunits come into play and the dimer is the important entity. The protofilaments that run lengthwise along microtubules are built from a head to tail alignment of tubulin dimers. Protofilaments associate laterally with a $\sim$0.9 nm offset so that the nearest neighbour monomers describe 12 nm pitch helical pathways when followed from protofilament to protofilament around the tube. It is convenient to classify microtubules by the two most prominent helical families that completely fill the monomer surface lattice. Thus, we use the nomenclature N : S to refer to a microtubule with a surface lattice defined by N protofilaments and S shallow-pitch, left-handed, monomer helices, see Fig. 1.

In vitro assembly of MAP free tubulin (PC-tubulin) usually gives a range of microtubule structures with protofilament numbers N in the range $10 \leq N \leq 16$ and values of S (the so-called start number) in the range $2 \leq S \leq 4$. Microtubule images obtained by electron cryomicroscopy have been the main source of understanding how microtubules can exist with a large range of protofilament numbers [17,102,103]. For protofilament numbers other than 13, the S start helices cannot fit together correctly unless the surface lattice mismatch is taken up by an overall rotation of the surface lattice. As a result the protofilaments are slightly skewed with respect to the axis of the microtubule and form long-pitch "superhelices"

[17,102]. Since the initial mismatch depends on both N and S, the pitch and the hand of the protofilament helices also depend on these parameters. For N in the range 11–15, the smallest lattice rotation, and hence the lowest energy paid for distorting the bond angles between tubulin subunits along the protofilament, is found for matching onto three start helices S = 3. However, other values of S are possible and unexpected values have also been observed for some protofilament numbers, for example 15 : 4. Overall, the lattice accommodation mechanism based on the monomer lattice gives an excellent description of the surface lattice organisation of microtubule polymorphs and argues in favour of "quasi-identical" interprotofilament contacts for both tubulin subunits.

In contrast, the dimer lattice is important when the interaction of microtubules and the motor proteins is considered. Current evidence is strongly in favour of the so-called B-lattice in which identical tubulin subunits are aligned along alternate shallow-pitch S-start helices [62,86]. For the B-lattice, S must be even for the surface lattice to have complete helical continuity [2,103]. Consequently the 13 : 3 microtubules usually found in eukaryotic cells have a packing discontinuity where the β-subunits aligned along a 3-start helix, switch after one complete turn to a set of α-subunits and so on from turn to turn. As a result, two protofilaments have neighbouring α and β subunits along the 3-start helix, this is the so-called "seam". It is not known whether the presence of this "defect" has any functional rôle but it is a clear indication of a non-helical assembly path for microtubules.

A long-standing problem is the relationship between the structural and dynamic polarity of microtubules. The now classical method of determining microtubule polarity in vitro involves growing C-shaped unclosed tubules onto cellular microtubules [38]. Unfortunately, this method cannot be applied to structural investigations of motor proteins interacting with in vitro assembled microtubules where it is clearly important to know the microtubule polarity. Careful examination of microtubules grown off the ends of sea urchin sperm flagella, using N-ethylmaleimide treated tubulin to block minus end growth, shows the main mass of the kinesin motor to be on the plus end oriented tubulin subunit [42]. An α-tubulin specific antibody attaches only to microtubule minus [23] and GTP-coated fluorescent beads interact only with the plus end [67]. Based on this evidence a convincing consensus is emerging in favour of a polarity assignment with β-tubulin at the fast growing microtubule plus end. In another approach, nucleation from centrosomes was used to obtain microtubules with a defined polarity and it was shown that the arrow-like moire fringes that characterise electron micrographs of vitreous ice embedded specimens point towards the microtubule plus end for microtubules with right-handed superhelical protofilaments and towards the minus end for left-handed protofilament helices [16]. This method can be used to determine the polarity of any individual microtubule observed by electron cryomicroscopy in vitreous ice.

4. The kinesin family of motor proteins

The kinesin superfamily of motor proteins use ATP hydrolysis to fuel movement along microtubules [12]. For example, *Drosophila* kinesin moves at about 1 μm/s along the protofilament direction towards the microtubule plus end [71] with a probable step length of around 8 nm [92]. This corresponds to the spacing between tubulin dimers along the protofilament leading to the idea that kinesin may move along a single protofilament or astride a pair of neighbouring protofilaments. Kinesin remains in contact with microtubules over many steps, a characteristic called processivity [9,28,33,49].

Since kinesin was first isolated in the mid 1980s, many other proteins have been discovered with similar sequences in a region of some 340 amino-acids, called the motor domain, that includes the ATP binding site, and the microtubule binding regions. There are now upwards of 90 members of the kinesin superfamily listed in sequence data bases. These proteins are omnipresent among eukaryotes where, in partnership with microtubules, they are involved in cell division, in intracellular transport and in the organisation of the cytoplasm. Similarly to myosin, there are a whole range different kinesins, so that the kinesin superfamily can be divided into sub-groups of kinesin-like proteins (KLPs) that may be monomers, hetero- or homo-dimers, trimers, tetramers [18]. In their "standard" form these proteins are elongated heterotetrameric proteins with two heavy and two light polypeptide chains with molecular weights typically in the ranges $\sim$110–130 kDa and $\sim$60–80 kDa, respectively. The heavy chains are organised into three distinct regions: the motor domain, a rod-like region and a globular tail [41,82]. The motor domain is usually situated at the amino-terminal end of the heavy chain and is followed by a region predicted to be α-helical with a 7-residue repeat (abcdefg) with hydrophobic residues dominant in positions a and d. Such sequences are predicted to favour heavy chain dimerisation by forming α-helical coiled-coils. The globular tail, in partnership with the light chain, is thought to be involved in the cargo specificity.

Like myosin, kinesin is an ATPase and ATP hydrolysis is highly stimulated by the presence of microtubules. Kinesin-ATP appears to have a much higher affinity for microtubules than does kinesin-ADP and consequently the synchronisation of the ATP hydrolysis cycle and the cycle of attachment and detachment is likely to be quite different to that of myosin. Most kinesins move towards the plus end of the microtubule but some, like ncd, move towards the minus end.

5. Structure of actin

Actin has been crystallised as a stable 1 : 1 complex with DNaseI, Fig. 2 [47,56]. The major role of DNaseI in the crystallisation protocol is to inhibit the self as-loop 41-50

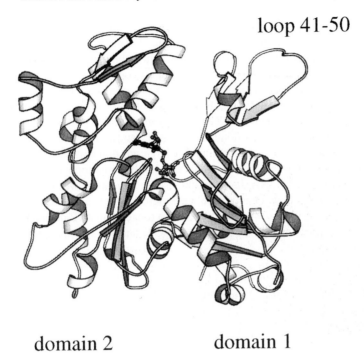

domain 2   domain 1

Fig. 2. The actin monomer consists of two main domains with ATP (shown as a ball and stick model) sandwiched in between [56]. The hydrophobic loop 41–50 is probably involved in contacts between actin monomers along actin filaments. Two other ATP binding proteins, hexokinase and the ATPase region of Hsc70 have very similar structures but very different sequences. This figure was built from the atomic coordinates reference 1ATN in the Protein Data Bank. Figs. 2 to 4 were made by F. Kozielski using the program MOLSCRIPT [58].

sembly of actin filaments. The structures of the ADP and ATP associated forms determined at 2.8 Å and 3.0 Å resolution show that actin has two similar domains each with five stranded β-sheets. Surprisingly the structures in the presence of the two nucleotides are very similar although this could possibly be because conformational changes are blocked by the presence of DNaseI, or by constraints imposed by the crystal packing. The domains are connected by two peptide loops near amino-acids 140 and 338 and are related by an approximate 2-fold axis. As shown in Fig. 2, the nucleotide is located between the domains and stabilises the structure via salt bridges and hydrogen bonds. Amino-acids 41–50 form a hydrophobic loop that interacts with DNaseI and that is expected to contact domain 2 in the next actin molecule in F-actin.

Oriented actin gels give X-ray fibre diffraction to around 8 Å and this allows the structure of actin filaments to be deduced by adjusting the atomic structure of the monomer into a model of F-actin to obtain the best fit between the observed and calculated diffraction intensities [46]. Comparable results are obtained by fitting the atomic model into the molecular envelopes obtained by three-dimensional helical reconstructions from electron cryomicroscopy images [66].

An unexpected and remarkable feature of the atomic structure of actin is the similarity to two other ATP binding proteins that have no sequence or functional homology, hexokinase and the ATPase fragment of Hsc70 [45]. All three proteins are organised into two main domains with ATP sandwiched between them. Each domain divides into two sub-domains. The large sub-domains Ia and IIa have five-stranded β-sheets joined by two α-helices with the same topography and related by a pseudo-two fold axis. In actin and Hsc70, the gap between domains I and II is closed whilst hexokinase has a wide cleft, leading to the idea that the two domains may be hinged and might be able to rotate in an ATP-dependant way. In actin the putative hinge points would be Gly 342 and Ala 144 at the bottom of the structure in Fig. 2.

6. The structure of myosin and kinesin

6.1. Low resolution structures

These proteins are members of two distinct and unrelated superfamilies each of which is characterised by a high degree of sequence homology in their function-ally important motor domain. The myosin heavy chains form dimeric molecules with two globular domains (S1) connected by a long α-helical coiled-coil. The S1 region has motor activity, i.e. it contains the ATP site and the regions that interact with filamentary actin. The essential and regulatory light chains interact with the S1 region of myosin. They are usually considered as essential for myosin to func-tion correctly. Kinesin heavy chains also form a two headed dimer with a long stalk organised as a coiled-coil. There is no sequence homology between kinesin and myosin. Kinesin light chains interact with the heavy chain "tail" regions that are distal to the globular heads. Most proteins belonging to the kinesin family have the motor domain at the N-terminal region of the heavy chain. There are some exceptions, like *Drosophila* ncd and Kar3 from *Saccharomyces cerevisiae* (baker's yeast), with the motor region at the C-terminus and it may be significant that ncd moves towards the microtubule minus end.

6.2. Structure of myosin

Myosin is abundant and readily purified from muscle. The globular head domain, sub-fragment1 (S1), can be prepared in large quantities by limited digestion using the proteolytic enzyme papain. Despite this, the first successful crystallisation of the 130 kDa S1 fragment from chicken skeletal muscle was only possible after reductive methylation of all the lysine residues [73,75,93]. The structure was solved to 2.8 Å resolution, 1072 residues out of 1157 were located, the missing residues are loops and the ends. There are three distinct β-sheet regions. The central motif is made up of a large almost parallel 7-stranded β-sheet, shown dark in Fig. 3, flanked on both sides by three α-helices. Strands 1 and 6 of this sheet are in the opposite orientation to the other five. In the centre there is a classical P-loop with the sequence GxxxxGK(S/T) that characterises ATP and GTP binding proteins such as adenylate kinase and Ras. In myosin, this phosphate binding loop GESGAGKT connects a β-strand to an α-helix and identifies the catalytic site. The nucleotide pocket is about 13 Å wide and deep. Overall, the structure of S1 myosin is 48% α-helical with a very long α-helix at the COOH-terminal end. In the native molecule this helix connects to the coiled-coil α-helix. The essential and regulatory light chains wrap around this helix and, amongst other effects, could rigidify the lever-arm. S1 has an asymmetric structure with approximate dimensions 165 Å by 65 Å by 40 Å and it can be proteolysed into three major fragments, the 25 kDa N-terminal region, the 50 kDa central region and the 20 kDa C-terminal region. The 50 kDa fragment has two major domains separated by a cleft at about 90 Å from the COOH-terminus. Closure of the cleft could move the end of the 85 Å long C-terminal α-helix by about 60 Å. The actin binding region is expected to be on the exposed loops containing hydrophobic residues at a distance of some 35 Å from, and on the opposite side of the central domain to the ATP binding pocket [31,76].

6.3. Structure of kinesin

It is difficult to obtain large amounts of pure kinesin from natural sources, for example brain tissue, and almost all recent work has used bacterial overexpression to obtain truncated proteins corresponding to specific regions of the heavy chain. This has allowed large amounts of protein to be obtained for crystallisation assays and the crystal structure of the motor domain some 340 amino-acids in length was obtained in 1996 for human kinesin [59] and for *Drosophila* ncd [78]. Compared to myosin S1, kinesin is very small, it's motor domain is only about 340 amino-acids in length, a mere third of the size of the S1 fragment. The structure of human kinesin expressed in *E. coli* has been determined to 1.8 Å with bound Mg-ADP. It is an arrow-shaped molecule 70 Å by 45 Å by 45 Å with amino-acids 7 to 325 visible in the structure. The core of the molecule is the seven stranded β-sheet, mostly parallel and shown dark in Fig. 4, with three α-helices on each side. The nucleotide cleft is open and exposed to the solvent. In view of the lack of sequence homology and of the large size difference compared to myosin it was

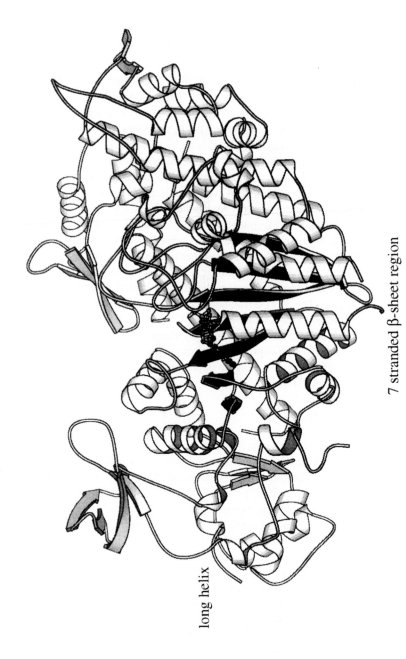

7 stranded β-sheet region

long helix

Fig. 3. The ATP binding region of the myosin S1 fragment is a large 7 stranded β-sheet central motif, dark colour, flanked by α-helices [75]. The putative actin binding regions are on the opposing face of the molecule. This figure was constructed from data referenced as 1MMG in the Protein Data Bank.

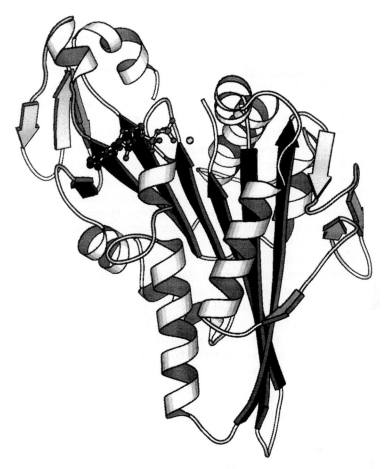

Fig. 4. The structure of the kinesin motor domain has an almost identical central core to that of myosin S1, a 7 stranded β-sheet with three α-helices on either side, again shown in dark colour [59,75]. There are no significant sequence similarities between the two proteins except for the loops involved in the binding of ATP. This figure was constructed from F.J. Kull's kinesin structure pages on the kinesin homepage.

a considerable surprise to find that kinesin and ncd show a strong structural similarity to the central core of the myosin motor domain, the β-sheet region with the associated α-helices on either side [59,72,83,85,99]. This led to suggestions that some elements of the tertiary structure of myosin may have structural and functional analogues in the kinesins. For example, the α-helical regions connecting to the coiled-coil, absent in the first kinesin and ncd crystal structures, were tentatively positioned by analogy to myosin's long α-helical lever [99]. Such predictions are delicate, as witnessed by comparison to the recent crystal structure of monomeric and dimeric rat kinesin that shows the neck region α-helix in a radically different position and orientation [57,79]. Also the big size difference between myosin S1 and kinesin is largely due to two long inserts in myosin that include the actin binding regions, so it was proposed that the equivalent, shorter, inserts in kinesin contain the microtubule interaction regions [59]. In kinesin these regions run from amino acids 138 to 173 and from 272 to 280. Apart from the long inserts, the topology and tertiary structure of kinesin is remarkably close to that of the myosin core; within seven of the core β-strands and six α-helices the α-carbon backbones of the two proteins fit with a r.m.s. difference of 3.5 Å over 183 amino-acids despite the fact there are no sequence similarities apart from the residues involved in the phosphate binding loops. Moreover six of the seven β-strands and the six α-helices run in the same direction. The nucleotide binding region in kinesin shows a classical P-loop (GxxxxGK(S/T)) and there are other loops near the γ-phosphate position that have the same orientation as in myosin and that are equivalent to the G-protein switch regions I (SSR, metal binding) and II (DLAGSE, phosphate sensitive glycine) [99].

The crystal structure of the ncd monomer to 2.5 Å resolution is very similar to kinesin despite the fact that it is located at the C-terminal end of the heavy chain and that this protein is a minus-end directed motor. Very recently the structures of the rat kinesin monomer and dimer and of Kar3, a minus end directed motor in yeast, have also been solved [30,57,79].

7. Structure of microtubule-kinesin and actin-myosin complexes

The crystal structures will be essential to an overall understanding of the motor proteins but by themselves they cannot tell us, for example, how microtubules and kinesin interact. This question has been addressed by electron cryomicroscopy for both systems. For actin-myosin see references [55,65,104]. We can illustrate the approach by discussing the microtubule-kinesin system. Kinesin and ncd monomers complexed to microtubules are found to bind with a stoichiometry of one monomer per tubulin dimer [35,86]. This specific decoration of the microtubule surface lattice has provided direct evidence that *in vitro* assembled microtubules have a B-lattice organisation. The three-dimensional reconstructions of microtubule-kinesin monomer complexes clearly show this lattice organisation and the 8 nm kinesin spacing corresponding to the tubulin dimer spacing along the protofilaments. Although the monomer complexes have given very informative results about the microtubule lattice and interaction stoichiometry, it is better to work with the functional dimers to obtain information relevant to motor movement, directionality and details of the protein–protein interactions between kinesins and microtubules.

Overexpressed recombinant kinesin heavy chains longer than some 380 amino acids spontaneously form dimers and are viable motors [33,105] these have been used in several investigations of microtubule-dimer complexes. *Drosophila* kinesin and ncd are usually used as archetype plus-end and minus-end directed motors, respectively. These proteins have been well characterised [14,52], and both binding assays and electron cryomicroscopy have indicated that, in the presence of the slowly hydrolyzable ATP analogue, AMP-PNP, these dimers interact with taxol-stabilised microtubules with a stoichiometry of one motor dimer per tubulin heterodimer [4,35,43]. Each of the two dimers has one attached, and one unattached head. The unattached heads have distinctly different conformations that might be related to their opposite directionality. Recently the relationship between the structure of microtubule-kinesin complexes and the nucleotide state of the kinesin dimer has been investigated. Images of complexes in the presence of ADP, ADP-AlF$_4$ (to mimic the ADP-P$_i$ state) and apyrase (to mimic the no nucleotide state) have been obtained [5]. Again, the kinesin dimer is always found to have one attached head and one free head. The free head contacts the end of the attached head closest to the microtubule plus end, and it points sideways to the right and upwards towards the microtubule plus end, i.e. in the direction of movement of the kinesin motor. The free head appears to occupy a smaller volume than the attached head and our current belief is that the apparent size difference between the two heads is due to positional disorder of the free head that is tethered to the attached head via the neck region and the coiled-coil.

Experience with the actin/myosin system [74], and with other protein complexes has shown that it is possible to achieve an accurate fit of atomic resolution structures (obtained by X-ray crystallography) within the 2 to 4 nm resolution three dimensional molecular envelopes reconstructed from micrographs of vitreous ice embedded specimens. It is clear that one of the essential steps towards understanding microtubule/molecular motor systems will involve the combination of 3D structures from X-ray crystallography with those from electron cryomicroscopy, and the comparison of these results with those available from other techniques.

8. Assaying molecular movement and forces

Although the structural work discussed above is essential in order to understand the molecular details of motor protein movement, a lot of additional information will be required to answer all of the questions related to how motor proteins actually move along their pathway; how far they move; do they step, glide or diffuse; are two heads necessary for movement; what force can they develop; how efficient are they; how is ATP hydrolysis coupled to movement; which parts of the molecules are essential for motility; does myosin and kinesin movement have common features?

Progress in optical microscopy including the use of fluorescent markers of microtubules and F-actin, video microscopy methods including video enhanced differential interference contrast microscopy (DIC), dark field microscopy, the development of optical tweezers, microneedles and so forth has led to a really amazing situation where it is now possible to directly observe the movement due to individual motors and to measure the forces that they exert. Indeed optical microscopy (DIC) played a major role leading to the discovery and isolation of kinesin [80].

Over the past few years a large number of light optical observations have been made on both the F-actin/myosin and the microtubule/kinesin systems. In general, it is easier to work with the microtubule/kinesin system mainly because, (i) microtubules are more rigid than actin and, (ii) kinesin moves long distances along microtubules before loosing contact. Essentially two types of experimental setup have been used to investigate the kinesin-microtubule system. As shown in Fig. 5 kinesin can be adsorbed to the glass wall of a thin chamber and attaches microtubules which move in the presence of ATP, alternatively microtubules can be attached to the glass substrate and kinesin "tagged" at the distal end with silica or polystyrene beads is injected into the chamber to interact with, and move along, the microtubules. The movement of the bead is observed by light microscopy. In some experiments the bead is trapped with optical tweezers and the movement measured by interference methods.

Using these methods, dimers were shown to move along microtubules for considerable distances before detaching whereas monomeric kinesin detaches frequently [9,19,33,49–51,100]. Individual kinesin dimers can move along microtubules at a speed of up to 1 μm/s and the speed appears to be independent of the number of kinesin molecules interacting with the microtubule. The movement is along the protofilament direction. This has been shown very clearly in experiments by J. Howard's laboratory in which microtubules with different numbers of protofilaments were shown to rotate as they moved along kinesin tethered to glass slides. The rotation was correlated to the superhelical pitch of the protofilaments as obtained by electron cryomicroscopy [71]. Tethered kinesin interacts with microtubules with a considerable degree of rotational freedom, rotations of well over 360° have been observed [53].

Work from the laboratory of S. Block using a differential interferometric method and optical trapping to overcome the Brownian movement of kinesin coated 1 μm silica beads interacting with tethered microtubules has shown that kinesin dimers move along microtubules with 8 nm steps [10,48,91,92]. By re-

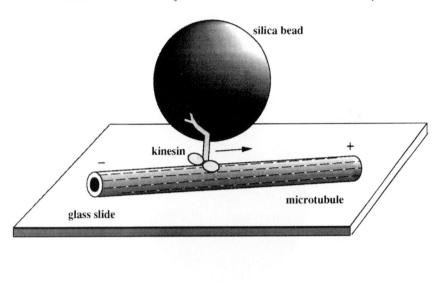

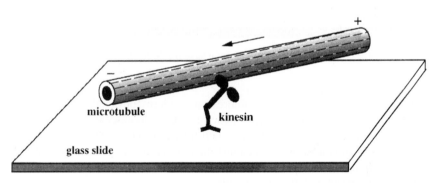

Fig. 5. Schematic drawings of experimental setups for kinesin motility assays. Above: microtubules are attached to a glass cover slip in an environmental chamber and interact with kinesin covered silica beads. Below: kinesin is attached to the cover slip and moves microtubules in the presence of ATP. Movement and force can be measured by a number of light optical methods.

straining the movement of kinesin, the optical trap can be used to measure the forces exerted during movement, or in an alternative method microtubules can be attached to thin glass rods that bend when the microtubule is moved by tethered kinesin molecules. The forces required to prevent kinesin moving are of the order 5 pN [64]. Recently it has been shown that at low loads each 8 nm advance of kinesin is accompanied by the hydrolysis of one ATP molecule [51,81].

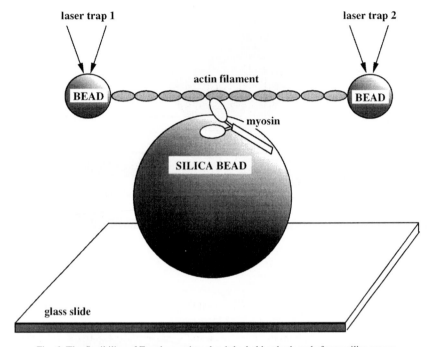

Fig. 6. The flexibility of F-actin requires that it be held at both ends for motility assays.

Using total internal reflection fluorescence microscopy, individual kinesin dimers with an attached C-terminal maleimide-Cy3 fluorescent dye have been shown to move some 600 nm along a microtubule before detaching [100]. This type of observation leads to the idea that the kinesin dimer advances by a 'hand over hand' or processive type of movement in which one or other of the two heads is always in contact with the microtubule.

In the case of the actin/myosin system the motility experiments are somewhat more complicated because, unlike kinesin, myosin spends much of its ATPase cycle detached from the actin filament and because F-actin is more flexible than microtubules [13,24,40,68,84,87,97,106]. The initial force measurements, made by attaching fluorescent actin filaments to a flexible glass needle, involved from 10 to 100 myosin heads interacting with the filament. The force was estimated as around 1 pN per myosin head. More recently another method has been used in which an actin filament is attached at both ends to silica beads using "inactivated" myosin. This actin filament can be stretched between two optical traps interacting with the silica beads as shown in the Fig. 6. This stretched filament can be manoeuvred to interact with myosin attached to another bead sitting on top of a microscope cover slip. Steps in the range 8–17 nm and forces of 3–7 pN were measured. As might be expected, thermal noise is a major problem in these experiments and it is important to find an adequate statistical approach to analyse the extremely noisy data sets. For example, in another investigation, taking account of small interactions "hidden" in the thermal noise, the measured step lengths were 4 nm. Isolated single and double headed myosins were found to exert similar forces of about 1–2 pN.

Some recent work has shown that single fluorescent molecules and individual ATP turnover by myosin in solution can be imaged by light microscopy based on total internal reflection [26]. This is a major advance compared to the most sensitive ATPase assays that require some 1000 myosin molecules, and perhaps the most important factor allowing the direct observation of single hydrolysis events is the extremely low level of background light at about 3 photons/s/resolution element.

9. How do motor molecules move?

Understanding how the motors work at a molecular level will involve a knowledge of how the cycles of movement and ATP hydrolysis are coupled to each other and to conformational changes in myosin and kinesin and perhaps also in their pathways, actin and tubulin [32,54,89]. In skeletal muscle the basic operating unit is the sarcomer in which thick (myosin) and thin (actin) filaments overlap and a large number of myosin motor domains interact with each thin filament. The situation is completely different for kinesin based axonal transport where it appears that individual kinesin molecules are involved in moving specific cargoes from the cell body to the synapse. Consequently it is likely that the functioning of the actin-myosin and the microtubule-kinesin systems may be rather different.

Myosin and kinesin are chemico-mechanical enzymes that use ATP hydrolysis to fuel movement. Their ATPase activity is strongly stimulated when they interact with their specific pathways. The basic hydrolysis cycle from ATP to ADP must take place during the period that the nucleotide is attached to the motor and is expected to involve the following successive steps:

$$\text{motor-ATP} \leftrightarrow \text{motor-ADP-P}_i \leftrightarrow \text{motor-ADP} \leftrightarrow \text{motor.}$$

However complicated the details might be, there are basically only three or four distinct phases in the movement of the motors along the pathway:

$$\text{attachment} \rightarrow \text{stroke} \rightarrow \text{detachment} \rightarrow \text{reverse stroke.}$$

Naturally, the question arises as to how the cycles of ATP hydrolysis and attachment–detachment are synchronised and as to how the conversion of chemical energy into movement is influenced by the interaction of the motor with the "pathway", and by the force required to move the cargo under different loads. Is movement based exclusively on conformational changes in the motor molecule as in the power stroke model or is the energy released by ATP hydrolysis used to bias random movements due to thermal agitation? Our understanding of these basic questions will improve as quantitative data on the simplified model systems discussed in the previous section is used to improve physical descriptions of motor function [6,20,21,50,60,70].

The power stroke molecular model of the contractile cycle for the actin/myosin system is based on two main ideas: (i) the direction of movement on the actin filament is defined by the polarity of both the actin and the myosin filaments, (ii) the ATP hydrolysis cycle induces conformational changes in the myosin molecule that modulate the affinity for actin. Since the myosin thick filaments are dipolar bundles of many individual myosin molecules a large number of heads are in close proximity to each actin filament in the sarcomer. To avoid frictional drag, an individual myosin head appears to attach to F-actin for only a short period during the complete cycle of ATP hydrolysis, i.e. it has a small duty cycle (duty cycle = attached time/unattached time), in the terminology of Leibler and Huse myosins are "rowers". For this motor it is unclear why the two heads are necessary for the motor function since two-headed myosin and the single motor domain S1 have both been shown to function in motility assays. According to the Lymn–Taylor cycle relating ATP hydrolysis and the swinging cross-bridge model, myosin-ATP is unattached, myosin-ADP-P_i attaches, the release of ADP and P_i induces a conformational change, ATP attaches and myosin releases.

Individual kinesin molecules are processive motors, i.e. they move along microtubules over large distances without "falling off". They are "porters" with a large duty cycle and probably need to always have one head attached to prevent them diffusing away from the microtubule. Kinesin-ATP, kinesin-ADP-P_i and kinesin alone all have strong affinities, kinesin-ADP has a weak affinity. Consequently, for kinesin, the motor-ADP state is likely to be unattached whereas for myosin it is a strong binding state.

Genetic engineering can play an important rôle in investigating the functional importance of various parts of the molecular motors. A straightforward example concerns the long α-helical region at the C terminal end of the S1 structure. If this helix is a lever involved in the power stroke, changing its length or rigidity should influence the step length and the overall velocity in motility assays. In vitro motility assays comparing mutants with shorter and longer necks to wild-type myosin has shown the sliding velocity to be proportional to the length of the lever [11,87,98]. Other possibilities involve the insertion of artificial lever arms. This has been done for the *D. discoideum* myosin S1 fragment by replacing the α-helical light chain binding domain by α-actinin repeats to give chimearic proteins with motility properties similar to the wild type protein [3]. Molecular genetics can also be used to investigate the functionality of randomly mutated forms of myosin.

In the case of *Drosophila* kinesin it has been shown that the direction of movement is an intrinsic property of the heavy chain motor domains. Fusion proteins with the α-helical tail region replaced at different positions by α-spectrin move towards microtubule plus ends like the wild type protein. The same result was found with GST fused to the N-terminus of the kinesin motor domain. However, two groups have recently shown the direction of movement can be reversed by fusing the kinesin stalk to the ncd motor domain and this seems to indicate that the "neck" region joining the motor domain to the coiled-coil region is involved in determining the direction of movement [13,39,90].

References

[1] L.A. Amos, Trends Cell Biol. **5** (1995) 48.

[2] L.A. Amos and A. Klug, J. Cell Sci. **14** (1974) 523. –549.

[3] M. Anson, M.A. Geeves, S.E. Kurzawa and D.J. Manstein, EMBO J. **15** (1996) 6069.

[4] I. Arnal, F. Metoz, S. DeBonis and R.H. Wade, Curr. Biol. **6** (1996) 1265.

[5] I. Arnal and R.H. Wade, Structure **6** (1998) 33.

[6] R.D. Astumian, Science **276** (1993) 917.

[7] K.B. Avraham, Nature **390** (1997) 559.

[8] M. Bähler, Curr. Opinion Cell Biol. **8** (1996) 18.

[9] E. Berliner, E.C. Young, K. Anderson, H.J. Mahtani and J. Gelles, Nature **373** (1995) 718.

[10] S.M. Block, Trends Cell Biol. **5** (1995) 169.

[11] S.M. Block, Cell **87** (1996) 151.

[12] G.S. Bloom and S.A. Endow, Protein Profile **2** (1995) 1105.

[13] R.B. Case, D.W. Pierce, N. Hom-Booher, C.L. Hart and R. Vale, Cell **90** (1997) 959.

[14] R. Chandra, E.D. Salmon, H.P. Erickson, A. Lockhart and S.A. Endow, J. Biol. Chem. **268** (1993) 9005.

[15] R.E. Cheney and M.S. Mooseker, Curr. Opinion Cell Biol. **4** (1992) 27.

[16] D. Chrétien, J.M. Kenney, S.D. Fuller and R.H. Wade, Structure **4** (1996) 1031.

[17] D. Chrétien and R.H. Wade, Biol. Cell **71** (1991) 161.

[18] D.G. Cole and J.M. Scholey, Trends Cell Biol. **5** (1995) 259.

[19] C.M. Coppin, J.T. Finer, J.A. Spudich and R.D. Vale, Proc. Natl. Acad. Sci. **93** (1996) 1913.

[20] I. Derenyi and T. Vicsek, Proc. Natl. Acad. Sci. **93** (1996) 6755.

[21] T. Duke and D.A. Leibler, Biophys. J. **71** (1996) 1235.

[22] P. Eldin, B. Cornillon, D. Mornet and J.J. Léger, Médecine/sciences **11** (1995) 1005.

[23] J. Fan, A.D. Griffith, A. Lockhart and R.A. Cross, J. Molec. Biol. **259** (1996) 325.

[24] J.T. Finer, R.M. Simmons and J.A. Spudich Nature **368** (1994) 113.

[25] S. Frankel and M.S. Mooseker Curr. Opinion Cell Biol. **8** (1996) 30.

[26] T. Funatsu, Y. Harada, M. Tokunaga, K. Saito and T. Yanagida, Nature **374** (1995) 555.

[27] F. Gibson, J. Walsh, P. Mburu, A. Varela, K.A. Brown, M. Antonio, K.W. Belsel, K.P. Steel and S.D.M. Brown, Nature **374** (1995) 62.

[28] S.P. Gilbert, M.R. Webb, M. Brune and K.A. Johnson, Nature **373** (1995) 671.

[29] H.V. Goodson and J.A. Spudich, Proc. Natl. Acad. Sci. **90** (1992) 659.
[30] A. Gulik, H. Song, S.A. Endow and I. Rayment, Biochem. (1998) in press.
[31] A.M. Gulick and I. Rayment, BioEssays **19** (1997) 561.
[32] D.D. Hackney, Annu. Rev. Physiol. **58** (1996) 731.
[33] D.D. Hackney, Nature **377** (1995) 448.
[34] J.A. Hammer III, J. Muscle Res. Cell Motility **15** (1994) 1.
[35] B.C. Harrison, S. P. Marchese-Ragona, S.P. Gilbert, N. Cheng, A.C. Steven and K.A. Johnson, Nature **362** (1993) 73.
[36] T. Hasson, P.G. Gillespie, J.A. Garcia, R.B. MacDonald, Y.D. Zhao, A.G. Yee, M.S. Mooseker and D.P. Corey, J. Cell Biol. **137** (1997) 1287.
[37] T. Hasson, M.B. Heintzelman, J. Santos-Sacchi, D.P. Corey and M.S. Mooseker, Proc. Natl. Acad. Sci. **92** (1995) 9815.
[38] S.R. Heidemann and J.R. McIntosh, Nature **286** (1980) 517.
[39] U. Henningsen and M. Schliwa, Nature **389** (1997) 93.
[40] H. Higuchi, E. Muto, Y. Inoue and T. Yanagida, Proc. Natl. Acad. Sci. **94** (1997) 4395.
[41] N. Hirokawa, K. Pfister, H. Yorifiyi, M.C. Wagner, S.T. Brady and G.S. Bloom, Cell **50** (1989) 867.
[42] K. Hirose, J. Fan and L.A. Amos, J. Mol Biol. **251** (1995) 329.
[43] K. Hirose, A. Lockhart, R.A. Cross and L.A. Amos, Proc. Natl. Acad. Sci. **93** (1996) 9539.
[44] K.C. Holmes, Curr. Biol. **7** (1997) R112.
[45] K.C. Holmes, C. Sander and A. Valencia, Trends Cell Biol. **3** (1993) 53.
[46] K.C. Holmes, D. Popp, W. Gebhard and W. Kabsch, Nature **347** (1990) 44.
[47] K.C. Holmes and W. Kabsch, Curr. Opinion Struct. Biol. **1** (1991) 270.
[48] J. Howard, Nature **365** (1995).
[49] J. Howard, Ann. Rev. Physiol. **58** (1996) 703.
[50] J. Howard, Nature **389** (1997) 561.
[51] W. Hua, E.C. Young, M.L. Fleming and J. Gelles, Nature **388** (1997) 390.
[52] T.-G. Huang, J. Suhan and D.H. Hackney, J. Biol. Chem. **269** (1994) 16502.
[53] A.J. Hunt, F. Gittes and J. Howard, Proc. Natl. Acad. Sci. **90** (1993) 11653.
[54] H.E. Huxley, Ann. Rev. Physiol. **58** (1996) 1.
[55] J.D. Jontes, E.M. Wilson-Kubalek and R.A. Milligan, Nature **378** (1995) 751.
[56] W. Kabsch, H.G. Mannherz, K. Suck, E.F. Pai and K.C. Holmes, Nature **347** (1990) 37.
[57] F. Kozielski, S. Sack, A. Marx, M. Thormählen, E. Schönbrunn, V. Biou, A. Thompson, E.-M. Mandelkow and E. Mandelkow, Cell **91** (1997) 985.
[58] P.J. Kraulis, J. Appl. Cystallogr. **24** (1991) 946–950).
[59] F.J. Kull, E.P. Sablin, R. Lau, R.J. Fletterick and R.D. Vale, Nature **380** (1996) 555.
[60] S. Leibler and D.A. Huse, J. Cell Biol. **121** (1993) 1357.
[61] S.K. Maciver, BioEssays **18** (1996) 179.
[62] E. Mandelkow, Y.-H. Song and E.-M. Mandelkow, Trends Cell Biol. **5** (1995) 262.
[63] V. Mermall, P.L. Post and M.S. Mooseker, Science **279** (1998) 527.
[64] E. Meyhöffer and J. Howard, Proc. Natl. Acad. Sci. **92** (1995) 574.
[65] R.A. Milligan and P.F. Flicker, J. Cell Biol. **105** (1987) 29.
[66] R.A. Milligan, M. Whittaker and D. Safer, Nature **348** (1990) 217.
[67] T.J. Mitchison, Science **261** (1993) 1044.
[68] J.E. Molloy, J.E. Burns, J. Kendrick-Jones, R.T. Tregear and D.C.S. White, Nature **378** (1995) 209.
[69] E. Nogales, S.G. Wolf and K.H. Downing, Nature **391** (1998) 199.
[70] C.S. Peskin and G. Oster, Biophys. J. **68** (1995) 202s.
[71] S. Ray, E. Meyhöffer, R.A. Milligan and J. Howard, J. Cell Biol. **121** (1993) 1083.

[72] I. Rayment, Structure **4** (1996) 501.
[73] I. Rayment and H.M. Holden, Trends Biol. Sci. **19** (1994) 129.
[74] I. Rayment, H.M. Holden, M. Whittaker, C.B. Yohn, M. Lorenz, K.C. Holmes and R.A. Milligan, Science **261** (1993) 58.
[75] I. Rayment, W.R. Rypniewski, K. Schmidt-Bäse, R. Smith, D.R. Tomchick and M.M. Benning, D.A. Winkelmann, G. Wesenberg and H.M. Holden, Science **261** (1993) 50.
[76] I. Rayment, C. Smith and R.G. Yount, Ann. Rev. Physiol. **58** (1996) 671.
[77] M.K. Reedy, Structure 1993 (1993) 1.
[78] E.P. Sablin, F.J. Kull, R. Cooke, R.D. Vale and R.J. Fletterick, Nature **380** (1996) 550.
[79] S. Sack, J. Müller, A. Marx, M. Thormählen, E.-M. Mandelkow, S.T. Brady and E. Mandelkow, Biochem. **36** (1997) 16155.
[80] E.D. Salmon, Trends Cell Biol. **5** (1995) 154.
[81] M.J. Schnitzer and S.M. Block, Nature **388** (1997) 386.
[82] J.Y. Scholey, J. Hense, J.T. Yang and L.S.B. Goldstein, Nature **338** (1989) 355.
[83] J.R. Sellers, J. Muscle Res. Cell Motility **17** (1996) 173.
[84] R. Simmons, Curr. Biol. **6** (1996) 392.
[85] C.A. Smith and I. Rayment, Biophys. J. **70** (1996) 1590.
[86] Y.-H. Song and E. Mandelkow, Proc. Natl. Acad. Sci. USA **90** (1993) 1671.
[87] J.A. Spudich, Nature **372** (1994) 515.
[88] J.M. Squire, Curr. Opinion Struct. Biol. **7** (1997) 247.
[89] J.M. Squire, J. Muscle Res. Cell Motility **15** (1994) 227.
[90] R.J. Stewart, J.P. Thaler and L.S.B. Goldstein, Proc. Natl. Acad. Sci. **90** (1993) 5209.
[91] K. Svoboda, P.P. Mitra and S.M. Block, Proc. Natl. Acad. Sci. **91** (1994) 11782.
[92] K. Svoboda, C.F. Schmidt, B.J. Schnapp and S.M. Block, Nature **365** (1993) 721.
[93] E.W. Taylor, Science **261** (1993) 35.
[94] M.A. Titus, Curr. Biol. **7** (1997) 301.
[95] K.M. Trybus, J. Muscle Res. Cell Motility **15** (1994) 587.
[96] K.M. Trybus, Y. Freyzon, L.Z. Faust and H.L. Sweeney, Proc. Natl. Acad. Sci. **94** (1997) 48.
[97] Y. Tsuda, H. Yasutake, A. Ishijima and T. Yanagida, Proc. Natl. Acad. Sci. **93** (1996) 12937.
[98] T.Q. Uyeda, P.D. Abramson and J.A. Spudich, Proc. Natl. Acad. Sci. **93** (1996) 4459.
[99] R.D. Vale, J. Cell Biol. **135** (1996) 291.
[100] R.D. Vale, T. Funatsu, D.W. Pierce, L. Romberg, Y. Harada and T. Yanagida, Nature **380** (1996) 451.
[101] K.L. Vikstrom and L.A. Leinwand, Curr. Opinion Cell Biol. **8** (1996) 97.
[102] R.H. Wade, D. Chrétien and D. Job, J. Molec. Biol. **212** (1990) 775.
[103] R.H. Wade and D. Chrétien, J. Struct. Biol. **110** (1993) 1.
[104] M. Whittaker, E.M. Wilson-Kubalek, J.E. Smith, L. Faust, R.A. Milligan and H.L. Sweeney, Nature **378** (1995) 748.
[105] E.C. Young, E. Berliner, H.K. Mahtani, B. Perez-Ramirez and J. Gelles, J. Biol. Chem. **270** (1995) 3926.
[106] T. Yanagida, Y. Harada and A. Ishijima, Trends Biol. Sci. **18** (1993) 319.

COURSE 3

THE CELLULAR MACHINERY FOR CHROMOSOME MOVEMENT

J. Richard McIntosh

Department of Molecular, Cellular and Developmental Biology, University of Colorado, Boulder, CO 80309-0347, USA

G. Zaccai, J. Massoulié and F. David, eds.
Les Houches, Session LXV, 1996
De la Cellule au Cerveau
From Cell to Brain: Intra- and Inter-Cellular Communication –
The Central Nervous System

47

Contents

49

1. Background ideas and facts

The growth and division of cells is fundamental for life. For unicellular organisms with true nuclei, such as algae, fungi, and protozoa, cell growth and division are obvious aspects of biological success: they comprise the mechanisms by which cells multiply their genomes. The vigor of such a species generally depends upon its ability to consume nutrients efficiently and convert them into the materials and chemical energy with which to increase its numbers relative to the competition. Cell division is also important for multicellular organisms. Most plants and animals begin their lives as a single fertilized ovum and develop into their adults forms through a series of cell divisions and differentiation processes. Moreover, cell division is important for the maintenance of adult tissues, since it is the process by which dying cells are replaced, and it is essential for the healing of wounds.

Not only the event of cell division is important; the quality of the process is critical for the health of an organism. Two aspects of quality are significant. First, the number of cell divisions must be appropriate. In a multicellular organism either too few or too many divisions will lead to disease. For examples, anemia results from a failure to make enough red blood cells, and both solid tumors and spreading cancers are manifestations of excess cell divisions. For a unicellular organism, too many or too few cell divisions will lead either to over-crowding or to inefficient utilization of available food resources.

The second important aspect of quality in cell division is the requirement that the job be done with precision. The two resulting "daughter" cells must be endowed with the information, materials, and energy necessary to grow and divide again and/or with which to differentiate properly; this requires that important cellular constituents be sorted appropriately as a cell divides. For example at every cell division, nuclear DNA, the cell's repository for genetic information, must be duplicated and segregated into two identical sets, so each daughter cell will receive a complete set of the instructions necessary to build a new cell. DNA replication occurs during *interphase*, the period of cell growth between two cell divisions. Also during interphase, cells must synthesize sufficient quantities of all their macromolecules (proteins, nucleic acids, lipids, and polysaccharides) to endow both daughter cells with the components they need to survive. Interphase is the time when the cell doubles biochemically, and the subsequent cell division is

51

the physical separation of a biochemically doubled cell into two distinct compartments. Several cellular objects essential for cell viability, e.g., the chromosomes, cannot be synthesized de novo, and their loss at cell division would be permanent and lethal; the division process must assure that no such mistake occurs.

Many cellular constituents, such as the ribosomes that synthesize protein, are present in thousands of copies per cell as the preparations for division are completed, so both daughters are highly likely to get approximately half of what is available. This is a simple consequence of the laws of large numbers, and no special segregation process is required. The same is true for essentially all the soluble molecules of the cell. Other cellular constituents are present in limited numbers. For example, in most animal cells there is only one *Golgi apparatus*, the cytoplasmic membrane system that modifies recently synthesized membranes and proteins by the addition and removal of sugars and marks them for transport to the correct cellular compartment. Prior to cell division, this apparatus fragments into pieces that are sufficiently numerous to be segregated by chance, and no special mechanism is necessary to assure its approximate equipartition.

Chromosomes, on the other hand, are present in only one copy per cell at the beginning of interphase and two after DNA replication is complete. The chances that the two cells arising from division will by chance get exactly one copy of each chromosome are minuscule. (For human cells with 46 chromosomes, each of which has duplicated for division, it is $2^{-92} \approx 7 \times 10^{-29}$). Meanwhile, the stakes for correct chromosome segregation are high; the loss or addition of even a single chromosome is often lethal. It is therefore no surprise that cells have built a special apparatus with which to accomplish accurate chromosome segregation. The segregation process is called *mitosis*, from the Greek word for thread, which describes the chromosomes as the cell prepares for division; the apparatus that moves the chromosomes is called the *mitotic spindle*. During the stage called *prophase*, which immediately precedes division, the chromosomes become visible in the light microscope as they *"condense"* by coiling. The word "chromosome" derives the fact that these objects bind strongly to many dyes and are thereby colored. As they become sufficiently condensed to be distinct, each chromosome appears double, reflecting the recent replication of the DNA and the fact that the two identical parts of the duplicated chromosome, called *chromatids*, are interconnected.

Chromosome compaction is essential for cell division because the DNA comprising each chromatid is very long and thin. All DNA double helices are about 2 nm in diameter, and the DNA in human chromatids, for example, ranges in length from 1–6 cm. During prophase, 92 such objects lie within the nucleus, a sphere whose diameter is generally $<8~\mu$m. It is obvious that without condensation, motions over distances of cellular dimensions (usually 10–50 μm) could not separate the duplicated genome into two distinct chromosome sets. Furthermore, the chromosomes do not generally occupy separate domains within the interphase nucleus. Their accurate segregation must therefore be preceded by a process that disentangles them and makes each chromatid small enough that its movement over cellular dimensions will separate it from its twin. The mechanisms for chromosome condensation are fascinating but lie outside the scope of this chapter. For a discussion of current facts and models, see [10] and [18].

This chapter will present a description of our current knowledge about the mitotic spindle, which binds to the chromosomes after they have condensed, organizes them within the cell, and then segregates them into two distinct sets. The spindle is a bipolar array of microtubules and their associated motor enzymes, cellular components described in the two previous chapters. The polymerization of spindle microtubules (MTs) from their constituent protein, tubulin, is initiated by two foci called *centrosomes* (which are described with more detail in the next chapter). MTs initiated by the centrosomes bind to the chromosomes and move them into an organized array at the stage called *metaphase*, during which the chromosomes lie at or near the midplane of the spindle. Cells possess a mechanism that identifies when adequate order has been established. They then initiate the spindle-dependent functions of separating the duplicated chromosomes and moving them to opposite ends of the cell. The quality control system that detects adequate order in the mitotic chromosomes is a fascinating example of cellular logic, but it too is beyond the scope of this article. The interested reader is referred to a recent review [34].

2. The events of mitosis

2.1. Chromosomes become attached to cytoplasmic microtubules

During interphase, the decondensed chromosomes are separated from the surrounding cytoplasm by the *nuclear envelope*, a barrier comprised of two membranes made of lipid and protein. The mitotic spindle usually begins to form in the cytoplasm during prophase. The primary constituents of the forming spindle are two centrosomes (formed by the duplication during interphase of the single centrosome that was inherited at the previous cell division). The centrosome is an animal cell's principal organizer of MT polymerization (reviewed in [8]). The active component of animal centrosomes is the *pericentirolar material* (PCM), a fibrous, darkly staining material that appears largely amorphous, though it is often arranged in a roughly spherical shell (Fig. 1a). PCM contains small assemblies of a few proteins that serve as seeds for the polymerization tubulin into MTs. These seeds are composed of a special form of tubulin (the γ-isoform) and several additional proteins that together can nucleate a MT with the 13 protofilaments that

J.R. McIntosh

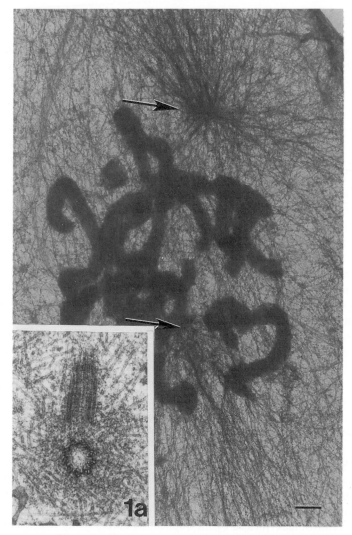

Fig. 1. Early prometaphase in a cultured mammalian cell, strain (PtK$_1$). (a) A thin section cut through a centrosome. Two *centrioles* are evident, the cylindrical structures oriented perpendicular to one another that are made from short MTs; the pericentriolar material (PCM) is arranged in a spherical shell around one of them. MTs emanating from the PCM are evident. (b) A high-voltage electron micrograph of a whole PtK cell stained with antibodies to tubulin, conjugated to 15 nm particles of colloidal gold to yield special contrast for the MTs. The nuclear envelope has just broken down, so the chromosome arrangement appears still to be roughly circular. Two centrosomes (arrows) are initiating arrays of MTs. These grow with approximately radial symmetry, so some MTs invade the area that used to be the nucleus. The bar = 1 μm.

comprise the wall of a cytoplasmic MT (reviewed in [38] and other chapters of this book). Tubulin is a dimer of two distinct subunits, α- and β-tubulin, which assemble head-to-tail, so each MT protofilament is structurally polar, with one end different from the other and with a vectorial character throughout its length (see the two previous chapters). Centrosomes define the polarity and geometry of a cell's MTs, as well as the number and position of the MTs that form (reviewed in [26]). In a mammalian cell, a mitotic centrosome will commonly initiate 200–400 MTs that project from the PCM with approximately radial symmetry. The β-tubulin ends project away from the centrosome [32], and since this is the MT end that can add and lose tubulin subunits more rapidly, it is also called the plus-end. The MT end associated with γ-tubulin at the PCM is comparatively static and is called the minus end.

As a cell enters mitosis, the two centrosomes initiate the MTs arrays described above, and the nuclear envelope disperses. Mitotic MTs can therefore grow into the region inhabited by the chromosomes (Fig. 1b). These MTs are highly dynamic and display an unusual form of assembly–disassembly steady state called *dynamic instability* (see chapter by Job). Under these conditions, the concentration of soluble tubulin is constant, while individual MTs undergo rapid changes in length. Most of the polymers are elongating, but at essentially random times, each MT passes through a transition (a catastrophe) to a state of rapid disassembly. The MT then shortens 10 to 100 times faster than it was previously elongating. During shortening, there is a modest probability of a second transition (a rescue), which returns the MT to its growth state. If a rescue does not occur, the γ-tubulin seed is presumably revealed, and it will soon initiate a new MT (reviewed in [17]).

A population of MTs displaying a dynamically unstable steady state is continuously consuming energy. The source of this energy is guanosine triphosphate (GTP), which is bound exchangeably to tubulin dimers. GTP, like its more commonly used counterpart, adenosine triphosphate (ATP), contains "anhydride" bonds, which will break upon the addition of a water molecule. Under physiological conditions, such *hydrolysis* yields one molecule of guanosine diphosphate (GDP) and one of *inorganic phosphate* (PO_4^{3-}) and releases $\sim$12 kcal of free energy per mole of GTP hydrolyzed (about 8×10^{-20} J/molecule or 0.5 eV, which is about $20 \times kT$ at room temperature). Soluble tubulin binds GTP more strongly than GDP, and since cells spend metabolic energy to maintain a higher concentration of GTP than GDP, most soluble tubulin is in the GTP-bound form. During polymerization, GTP is hydrolyzed to GDP, which remains bound to the polymerized tubulin, while inorganic phosphate is released into solution. (The exact time of this hydrolysis relative to subunit addition has not yet been determined.) A MT is therefore composed largely of GDP-tubulin, save small domains of GTP-tubulin that reside at its growing ends (reviewed in [7]).

Ironically, the GDP-tubulin of which MTs are composed is incapable of polymerization; even at very high concentrations and in the presence of MT seeds, it will not assemble. A MT is therefore highly unstable: whenever GDP-tubulin can find a root by which to leave the polymer, it will do so. Depolymerization by subunit loss from the lateral surface of the MT wall has never been observed, so the GTP-tubulin at the MT end probably serves as a cap to keep the MT stable most of the time. The occasional loss of this cap is likely to be the event that accompanies a catastrophe, leading to the rapid shortening of the MT. Rescue probably involves the chance addition of enough GTP-tubulin at the end of a shrinking MT to re-establish the cap and allow polymerization to recommence.

These considerations are important during mitosis for two reasons. Since the chromosomes and the centrosomes are near to one another at the time of nuclear envelope breakdown, the dynamic instability of centrosome-initiated MTs assures that the chromosomes will be peppered by growing MTs; over the course of a few minutes, most regions on a chromosome will experience the intrusion of a growing MT (Fig. 1b). Each chromosome possesses two specializations, called *kinetochores*, that can bind strongly to the β-tubulin end of a MT (Fig. 2). Dynamic instability endows a population of randomly growing and shrinking MTs with the ability to search space until they find the kinetochores that are essentially randomly distributed in the volume previously occupied by the nucleus. Any MT that grows into a kinetochore is stabilized by kinetochore binding, while MTs that fail to do so are still subject to dynamic instability. The likelihood that centrosome-initiated MTs will ultimately find kinetochores is thereby increased [11]. In short, the behavior of MTs is appropriate to give a very high probability that every kinetochore will wind up with MTs bound to it.

A second reason why the dynamic instability of MTs is important for mitosis is that later in the process, when the chromatids segregate, the MTs attached to them must shorten; during this movement, mechanical work is done on the chromosomes. The inability of GDP-tubulin to polymerize implies that energy is stored in an assembled MT, and this energy may contribute to the work done during chromosome movement. The source of this energy is ultimately the GTP that is hydrolyzed during MT polymerization, but the energy that is released during tubulin disassembly is probably derived from the conformational strain that is imposed on GDP-tubulin when it is assembled in the MT lattice. A strained configuration for GDP-tubulin in MTs is suggested by two lines of evidence: (1) During MT polymerization, the rate constants for the hydrolysis of GTP bound to tubulin, together with the release of its inorganic phosphate, are essentially equal to the rate of rephosphorylation of tubulin-bound GDP, showing that the energy of this GTP hydrolysis is not released as tubulin assembles [4]. (2) When MTs are depolymerizing, the tips of the shortening protofilaments bend out from the MT wall, curving in a plane that contains both the protofilament and the MT axis

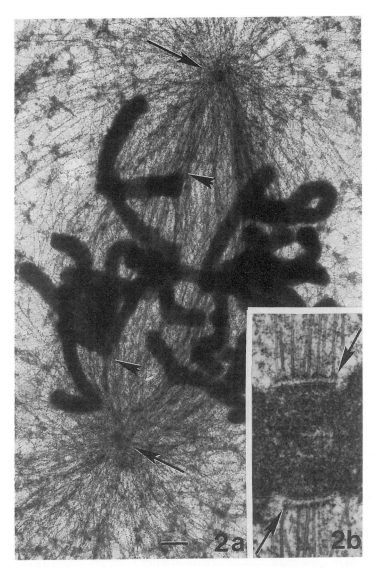

Fig. 2. Late prometaphase in a PtK cell showing the chromosomes already attached to the spindle and beginning to migrate to the spindle equator. In (a) many MTs are evident, and most of them run through the spindle-shaped region situated between the two centrosomes (arrows). Some MTs appear to end on chromosomes (arrowheads), and others pass the chromosomes to form an interdigitating framework that connects the two centrosomes. Bar = 1 μm. (b) A thin section of a prometaphase chromosome, showing closely associated sister chromatids, each of which has a kinetochore (arrows). Each kinetochores lies at the end of a bundle of MTs.

[24]. Such a curvature is incompatible with assembly into the cylindrical wall of the MT, and although curved protofilaments are seen only transiently during the depolymerization process, their occurrence suggests that bonds between adjacent protofilaments in the MT wall constrain the natural curvature of GDP-tubulin.

An upper bound for the amount of the energy stored per unit length of MT can be estimated from the energy of the one GTP hydrolyzed per tubulin polymerized. A tubulin dimer is 8 nm long, and 13 such dimers are assembled on a helical lattice. If all the energy of GTP hydrolysis were available to do mechanical work, the disassembly of one tubulin molecule could generate about 8×10^{-20} J/(8/13 nm), or 130 pN, which is more than $30\times$ the force generated by a single motor enzyme. Perfect efficiency is not, of course, expected, so the amount of force really generated would be less, but recent data, described below, suggest that such a process is actually at work during mitosis. The energetics of MT polymerization are therefore likely to be important for the energetics of chromosome motion.

2.2. Chromosomes become arranged so that each chromatid is attached to one centrosome

The most significant task in establishing the metaphase chromosome array is to insure that the two kinetochores on each chromosome (one of which is associated with each chromatid) become attached to MTs that are growing from different centrosomes. Put another way, high fidelity chromosome segregation is dependent upon assuring that "sister kinetochores" attach to "sister centrosomes" (Fig. 2). An important part of the mechanism by which this arrangement is achieved has been identified: when the attachment between a MT and a kinetochore or centrosome is under tension, it becomes stable. This somewhat surprising property was identified by a combination of descriptive and experimental cell biology. Occasionally during cell division, sister kinetochores attach to MTs growing from the same centrosome; the outcome of this aberrant situation has been followed in live cells by light microscopy. (Such aberrant arrangements are more common in the special cell division, called *meiosis*, that precedes the formation of egg and sperm cells.) This arrangement is unstable, and one or the other of the kinetochores loses its attachment to the nearby MTs, allowing the chromosome to attach to different MTs (reviewed in [33]). If the second attachment is again inappropriate (meaning that the MTs again come from the centrosome that is already attached to the sister kinetochore), it is again unstable, and the chromosome will again release its MTs, allowing it to try again, etc., until the appropriate orientation is achieved.

MT attachment to a kinetochore often leads to a brief, centrosome-directed movement of that chromatid [2], suggesting that MTs exert a centrosome-directed force on the kinetochore. This suggestion has been supported by surgical experiments with microbeams of light: when the regions of a chromosome that lie near a kinetochore are destroyed by a highly converging beam of either 260 nm light (which is absorbed by the chemical components of DNA) or by a pulse from a powerful infrared laser (e.g., neodymium-YAG), the kinetochore is detached from the rest of the chromosome; it then moves toward the centrosome to which it is attached by MTs [29,44].

Bruce Nicklas realized that forces from sister poles pulling on sister kinetochores that are still attached to one another, as they are before chromatid segregation, would put a chromosome under tension. He surmised that this tension might provide stability to the MT-kinetochore (and MT-centrosome) attachments. To test this hypothesis, he used fine glass needles to impale a chromosome that was inappropriately attached and exert the tension that would normally have come from the sister kinetochore's attachment to the sister centrosome. Experimentally applied tension imparted stability to the usually unstable, inappropriate attachment [3]. This manipulated chromosome never rearranged but went on to subsequent chromosome movements with both chromatids moving to one centrosome. Thanks to this and related experiments, it is widely thought that initial chromosome–centrosome attachments are largely random, but only the attachments that will lead to accurate segregation of the duplicate chromosomes to sister centrosomes are stable, as a result of the tension that they generate. Therefore, only the biologically appropriate attachments persist. Again, as with the attachments of dynamically unstable MTs to kinetochores, stable states are selected, because unstable states are, by definition, short lived. The structure that is biologically advantageous predominates because all other structures are unstable.

The mechanism by which tension promotes the stability of chromosome attachment is still a matter for speculation, but it may depend upon the decreased ability of motor enzymes to release from their associated fiber (e.g., a MT) when they are blocked in the completion of their power stroke by opposing forces [28]. Chemists describe this situation with *strain-dependent rate constants*.

If the sister kinetochores on unseparated chromosomes are being pulled towards sister centrosomes, then something must provide a counter force to keep the spindle from collapsing. Part of this force comes from centrosome-initiated MTs that do not attach to chromosomes but interdigitate with their counterparts from the opposite centrosome (Fig. 2a). Microbeams of 280 nm light have been used to destroy such MTs in some species, and the distance between the centrosomes then promptly decreases [21], suggesting that a centrosome-to-centrosome structure is in part responsible for maintaining spindle integrity. There are also indications from the behavior of spindles in yeast cells carrying mutations in cytoplasmic, MT-dependent motor enzymes that the centrosomes are pulled apart by forces generated outside the spindle [43]. These forces probably act on the *astral* MTs that project from the centrosomes into the surrounding cytoplasm (Figs. 2a and 3).

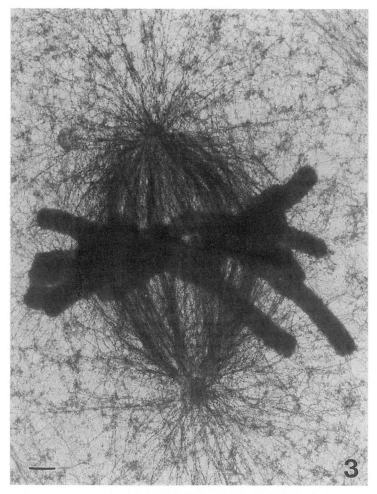

Fig. 3. A PtK cell in metaphase, showing the equatorial distribution of chromosomes in the fully formed mitotic spindle. The majority of the MTs are within the spindle itself, but a few project out into the surrounding cytoplasm, forming asters that are probably important for spindle stability and position. Bar = 1 μm.

2.3. Chromosomes become arranged at the equator of the spindle

The next major event of mitosis is the migration of the duplicate chromosomes, while they are still attached to both centrosomes, to the plane that lies equidistant between the centrosomes; this period of mitosis is called *prometaphase*. It leads to the geometry that gives the spindle its name and initiates the stage called *metaphase* (Fig. 3). These chromosome motions are highly irregular, showing an almost oscillatory path along circular arcs of various radii that pass through both centrosomes. The amplitude of the oscillations is only a few micrometers, about 1/5 the distance between the centrosomes, and their periods are many seconds long. The speeds of the chromosomes during these oscillations start at as much as 25 μm/min, but as prometaphase proceeds, they slow to as little as 1 μm/min. As the chromosomes follow these oscillatory paths, they gradually drift toward the spindle midplane, or *equator*, which lies equidistant between the ends, or *poles*, of the spindle.

The mechanisms for generating and controlling the forces for prometaphase chromosome oscillations and their drift toward the spindle equator are not yet fully understood. Certainly they involve the lengthening and shortening of the kineto-chore-associated MTs. Studies with labeled tubulin injected into living cells indi-cate that the majority of the subunit exchange that occurs during these movements takes place at the kinetochore ends of these MTs [54], but some MT depolymer-ization also occurs near the centrosomes. Indeed, throughout mitosis, the tubulin in kinetochore MTs is migrating toward the centrosomes at about 0.5 μm/min [31]. It is therefore not possible to say whether prometaphase chromosome mo-tion to the spindle equator is the result of forces generated at the kinetochores, the centrosomes, or along the fibers between them; there is some evidence for each position and probably all are true to some extent (reviewed in [26]). There is, however, a school of thought which argues that all the above forces are of minor importance, and that the motion of chromosomes to the spindle equator results from a pushing away from the centrosome that acts on the chromosome arms [39]. Hypothetically, this force is greatest near the centrosomes and weakens with distance, so chromosomes are forced to the spindle equator. While there is some evidence for this proposal, e.g., the presence of MT-dependent motor enzymes on chromosome arms [1,51,53], a serious problem for the proposal is found in the behavior of mitotic plant cells, whose chromosomes move nicely to the spindle equator, while the forces acting on the chromosome arms push *toward* the spin-dle poles, not away from them. Understanding prometaphase will require further work.

2.4. Each chromosome separates into two identical parts and these move apart

Once all the chromosomes are properly bound to spindle MTs, such that sister kinetochores are attached to sister poles, a clock is started [40]; within a few min-utes, sister kinetochores separate, allowing the duplicate chromosomes to sep-arate. The processes that detect the completion of chromosome attachment and the signals that initiate chromosome segregation are now under intensive study.

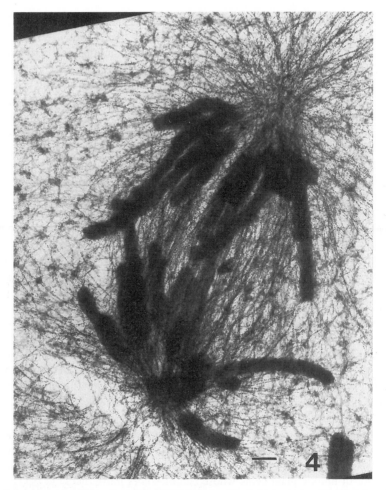

Fig. 4. A mid anaphase PtK cell, showing the already segregated chromosomes gathered at the poles; spindle elongation is just beginning. Some of the MTs that run between the poles have now formed bundles, and the astral MTs that project out of the spindle have already begun to elongate. Such MTs may play a role in placing the furrow that will form to divide the cell in two (see Fig. 5). Bar = 1 μm.

A plausible model has been proposed [25], and genetic analyses are identifying components of this mitotic *"checkpoint"* pathway (e.g., [41]).

The segregation of sister chromatids and their motion to the spindle poles is called *anaphase A* (Fig. 4). This movement is slow, usually about 1 μm/min. The overall impression from watching a time-laps display of anaphase is that chromosome motions are concerted and uniform, but careful examination has shown that individual anaphase chromosomes can stop and move backwards toward the spindle equator, then start poleward again. In this way anaphase A resembles prometaphase, but the net movement is in the opposite direction (it is now toward the spindle poles), and it is far less oscillatory.

The forces that would be required to move chromosomes against the viscous drag of cytoplasm at the speeds characteristic of anaphase A have been estimated by several workers at $\sim$0.01 pN, about 1/100 of the force generated by a single motor enzyme. The force necessary to stop a chromosome, however, exceeds this value by $\sim 10^5$, showing that something other than viscous drag is limiting the rate at which chromosome move [33]. Since MTs must depolymerize as chromosomes approach the spindle poles, it has been proposed that tubulin dynamics serve as a governor for the rate and extent of chromosome motions [27,33].

Shortly after the onset of anaphase A, the distance between the centrosomes begins to increase, a process called *anaphase B*. Descriptive work has shown that anaphase B, like A, proceeds at about 1 μm/min, and it depends upon both a sliding apart of interdigitating MTs and an elongating of these MTs by the addition of tubulin at their pole-distal ends (reviewed in [26]). These motions increase the final separation of the duplicate chromosome sets and make it comparatively easy for some simple mechanism, such as the furrow that develops at the equator of an animal cell, to separate the sister chromosome sets into two distinct cells (Fig. 5 and see also the chapter by Margolis). The division of one cell into two is called *cytokinesis*; generally it follows mitosis. Meanwhile, the nuclear envelope reforms around the separated chromosome sets, re-establishing the nuclei that are characteristic of interphase.

3. The mechanisms for chromosome movement

3.1. Motor enzymes, as well as microtubules, are important for chromosome movement

It is evident from the above that the formation of the mitotic spindle depends upon the controlled assembly of MTs, the capture of such polymers by kinetochores, and the subsequent depolymerization of kinetochore MTs, while the interpolar MTs elongate. For some years it was thought that these processes alone might be sufficient to provide the forces and organization for chromosome movement [15]. In many other biological situations, however, MTs are known to interact with motor enzymes, which use the chemical energy stored in ATP to generate forces and do work for movement along MT surfaces (reviewed in [52]). Two classes of MT-dependent enzymes are now known to be important for mitosis: cytoplasmic dynein and kinesin-like motors.

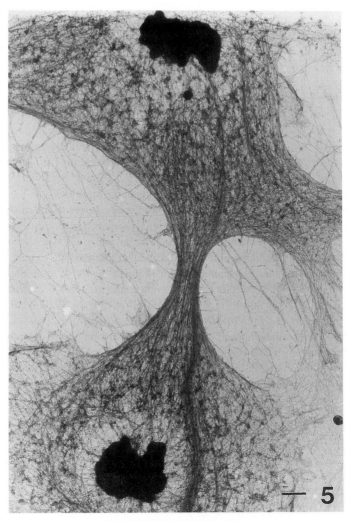

Fig. 5. A PtK cell that has completed anaphase and is in the process of cytokinesis. The cleavage furrow has decreased the diameter of the cell where the spindle equator used to be, and the residue of the spindle is now confined to the isthmus that connects the two cells. Nuclear envelopes are in the process of reforming to establish the interphase condition in which macromolecular synthesis can start again. Bar = 1 μm.

The enzymologies of both these classes of motors are quite well defined [9, 16], and particularly the kinesins are coming to be understood in molecular detail [42] (and see the chapter by Wade). These enzymes function as dimers of two identical polypeptide chains, each of which binds ATP and catalyses its hydrolysis to ADP plus PO_4^{3-}, which will remain bound to the protein for many seconds unless the motor encounters a MT. In that case, one of the two kinesin subunits binds to GDP-tubulin in the MT, catalyzing the release of first PO_4^{3-} and then ADP. During these release events, the enzyme changes it shape, likely shifting its position so the unbound kinesin subunit in the dimer is positioned to bind at some distance along the MT. Binding of ATP to the MT-associated kinesin subunit now promotes its release from the MT wall, and the cycle can be repeated. The distance most frequently traveled in such a step is 8 nm [47], the distance from one tubulin subunit to the next along a protofilament. A force of 4–5 pN will prevent the motion of kinesin along the MT, but when the load is less than this, there is a roughly linear relationship between force and velocity until kinesin reaches a speed of about 1 μm/s at under zero load at room temperature [13,46]. Velocity is also dependent upon temperature and the concentration of ATP.

Shortly after the purification of kinesin from embryos of the fruit fly, *Drosophila melanogaster*, the gene encoding its catalytic subunit was cloned [57]. This sequence was extremely informative, because it allowed the realization that several already-discovered mutations that perturbed mitosis in various fungi were changes in genes encoding kinesin-like molecules of these organisms [6,30]. Over the last ten years, a large family of kinesin-like proteins has been identified through a combination of biochemical, genetic, and molecular approaches. A simple yeast cell contains six kinesin-like proteins, five of which appear to play roles in chromosome movement [12]. Cytoplasmic dynein, too, contributes to the mitotic processes [49,58]. It is therefore evident that mitosis is not simply dependent upon MTs and their disassembly but also upon motor enzymes that can interact with MTs in rather complex ways.

Students of mitosis in fungi and animal cells have used both genetic and biochemical perturbations to determine the results of the failure or absence of a particular motor enzyme. Ironically, in the best studied example, the budding yeast *Saccharomyces cerevisiae*, no single motor enzyme is essential. Any one, and sometimes two or three such enzymes, may be eliminated by mutation without serious disturbance of chromosome segregation (reviewed in [12]). When multiple motors are mutated, however, the mitotic process fails, demonstrating the importance of motor enzymes for chromosome movement. These results suggest that the functions of mitotic motor enzymes are overlapping, so single motors can be missing and the mitotic machinery will still function (at least under laboratory conditions). This situation has led to the speculation that the system is over-determined to ensure reliability. Current work is designed to identify exactly where each motor enzyme is localized within the spindle and to learn just what process it accomplishes.

There is considerable evidence that kinesin-like motor enzymes contribute to the sliding of MTs that is part of anaphase B. A motor enzyme that localizes to the place where spindle MTs interdigitate has been studied in vitro; it will bind antiparallel MTs to one another in the absence of ATP and cause them to slide over one another in its presence [35]. Moreover, mutations in several kinesin-like proteins of yeast cause failure in the formation of a bi-polar spindle, and with temperature sensitive forms of the enzymes, together with the right combination of mutations in other genes, one can block spindle elongation [43]. Thus, kinesin-like proteins are strongly implicated as part of the mechanism for spindle elongation. Some data suggest that dynein contributes to anaphase B by anchoring in the cytoplasm and pulling on astral MTs that project out of the spindle, pointing away from the metaphase plate [43].

Significant attention has also been focused upon kinetochores, since these are obviously important parts of the machinery for chromosome segregation (reviewed in [37]). Kinetochores of *S. cerevisiae* have been investigated with special intensity, but although the DNA sequences for the kinetochore-associated parts of the chromosomes are known, and 4 proteins that form a complex to bind this DNA have been identified [19,20], the relevant motors have not yet been identified with certainty, and the mechanism by which the known proteins contribute to chromosome motion remains elusive. In vertebrate cells, kinetochores include at least three motor enzymes. One is a kinesin-like protein called CENP-E [59], which has the ability to walk on MTs from their plus ends towards their minus ends, i.e., from the kinetochore towards the centrosome [48]. The localization of this motor enzyme at the kinetochore is suggestive that it may play a role in chromosome-to-pole movement. Two subunits of cytoplasmic dynein have also been localized to kinetochores on vertebrate chromosomes [36,45], and a dynein-associated protein complex called "dynactin" (reviewed in [50]) has been found on kinetochores [5]. Dynein, like CENP-E, is a centrosome-directed motor enzyme, so it too is situated at the "scene of the crime", and may contribute to anaphase A. A third motor enzyme on mammalian kinetochores is MCAK, a kinesin-like protein whose polarity of motion has not yet been determined [55].

The multiplicity of motors at kinetochores may be due in part to the several functions that are accomplished by this chromosomal specialization. For example, kinetochores must capture MTs during the early stages of spindle formation. Motor enzymes attached to fibers that extend out from the kinetochore may increase the "scattering cross-section" of a kinetochore and thereby increase the probability of MT capture. Dynein appears to be so localized [56] and may play this role. Additional kinetochore functions include the motion of chromosomes both towards and away from centrosomes as they migrate to the spindle equator. One of the kinesin-like proteins of the kinetochore has centrosome-directed activity, but the other may work in the opposite direction. Bi-directional motion of MTs associated with kinetochores on isolated chromosomes has been observed under experimentally controlled conditions [14], but the molecular identities of these motor activities are not known, and the relevance of the motions seen in vitro to those occurring in a living cell remains to be determined. The factors that regulate these motors and assure both the metaphase configuration and anaphase onset are fascinating problems for future work.

3.2. Spindle motor enzymes can transduce the energy in assembled MTs into mechanical work to move chromosomes

There is direct evidence that CENP-E plays a role in chromosome movement on MTs during mitosis. Isolated chromosomes will attach to the plus ends of MTs in vitro, and when these MTs are induced to shorten, simply by dilution of the soluble tubulin, the chromosomes retain their attachment to the MTs as they depolymerize [22]. Surprisingly, this reaction will occur in the absence of added ATP, suggesting that the normal biological fuel is not necessary for the movement. Indeed, enzymes that degrade both ATP and GTP can be added to the preparation, driving soluble ATP concentrations below 10^{-10} M, and the velocity of chromosome movement on disassembling MTs is unaffected. We conclude that the energy stored in polymerized tubulin can be used to drive movements. The addition of antibodies to certain parts of the CENP-E molecule blocked this movement [22], suggesting that this motor enzyme plays a key role in coupling kinetochores to MTs in such a way that tubulin depolymerization can do work.

This activity for a motor enzyme is surprising, since the very essence of such enzymes is their ability to transduce the energy from ATP hydrolysis into mechanical work. We have therefore explored the phenomenon of MT disassembly-dependent motility in more detail. When kinesin is coupled to a 1 μm latex microsphere, this complex will bind to MTs; when ATP is added, the motors move the microspheres over MTs towards their plus ends. When tubulin is diluted, so the MTs disassemble, the kinesin coated beads are pulled towards the MT minus end, even when ATP is present [23]. These results suggest that motor enzymes can be forced "backwards" by tubulin disassembly.

This enigmatic reaction may be a subtle form of biased diffusion, similar to that discussed in the chapter by Job. Kinesin binds strongly to MTs in the absence of ATP. Under similar conditions, kinesin binding to soluble tubulin is weak, suggesting that tubulin assembly involves a change in the structure of the subunit itself. Consistent with this inference, disassembling MTs visualized by cryo-electron microscopy of fast-frozen polymers display a coiling of their protofilaments as subunits are lost from their ends (see above). The protofilaments appear to bend back, much as a banana is peeled [24]. Protofilament bending as part of the pathway for disassembly of GDP-tubulin offers a plausible mechanism by which the energy stored in the GDP-tubulin polymer could be released through conformational changes, altering the surface of the tubulin molecule and prying loose any motor enzyme that was bound while the tubulin was in the straight protofilament configuration that is characteristic of the MT wall. When motors situated near the MT end are pried off by this disassembly-associated conformational change, they may reposition by diffusion, binding to the remaining straight portions of the MT lattice. Since straight protofilaments lie on only one side of the motor, the region where disassembly has not yet occurred, motor diffusion is biased by movement of the disassembling MT end.

A disassembly-dependent mechanism clearly functions *in vitro*, but it is uncertain whether anything similar goes on within the living cell. Cells use many biochemical tricks to maintain millimolar concentrations of ATP, so it is implausible that a healthy cell would ever be without this energy source. Nonetheless, I propose that the disassembly-dependent reactions described *in vitro* comprise one part of the complex reaction system that is characteristic of MT-kinetochore interactions in vivo. For example, when a chromosome is moving to the metaphase plate, the MTs associated with one kinetochore must shorten while those bound to the sister kinetochore lengthen. The reactions we have seen *in vitro* may comprise one side of the more complex reaction system in which forces generated by motor enzymes and/or by disassembly are conjugated into a single reaction system by the molecules that bind kinetochores to MTs. Likewise, when a chromosome moves poleward in anaphase, not only must mechanical work be done to pull the chromosome through the cytoplasm, but MTs must depolymerize, losing subunits largely at their kinetochore associated ends. In short, the coupling of movement with MT disassembly may be the mechanism by which MTs serve as governors on chromosome motion.

4. Summary and conclusion

The above sketch of mitotic processes and mechanisms is an example of the complexity and elegance of an important biological event. Chromosome segregation is critical for cell division, so it is not surprising that considerable attention is devoted to accomplishing the process with fidelity. What has been particularly rewarding for students of the subject has been to watch so many layers of complexity emerge as the tools available for the study of cellular reactions have become more powerful. The "dance of the chromosomes" was discovered by structural studies with the light microscope more than 100 years ago. More recent work has revealed the remarkable MT behavior that accompanies it. Genetic and biochemical studies have now identified motor enzymes that associate with spindle MTs, and we are just beginning to identify the motor-associated proteins that probably help to connect these machines to the specific objects that must be moved to accomplish the cell's demands. One can readily imagine additional layers of control on motor velocity, force, and direction. The molecules involved in such control will emerge as our analysis of mitotic mechanism deepens. The important point that follows from our current understanding is that the molecules of the mitotic spindle, which seem to be as complex as a symphony orchestra, can be organized and controlled to produce not only symmetric machinery but highly precise events upon which the health of future generations can and will depend.

Acknowledgements

Work described here that was carried out in our lab at Boulder was supported in part by grants from the NIH (GM36663, GM33787, RR00592). JRM is a Research Professor of the American Cancer Society.

References

[1] K. Afshar, N.R. Barton, R.S. Hawley and L.S. Goldstein, Cell **81** (1995) 129.
[2] S.P. Alexander and C.L. Rieder, J. Cell Biol. **113** (1991) 805.
[3] J.G. Ault and R.B. Nicklas, Chromosoma **98** (1989) 33
[4] M. Caplow, R.L. Ruhlen and J. Shanks, J. Cell Biol. **127** (1994) 779.
[5] C.J. Echeverri, B.M. Paschal, K.T. Vaughan and R.B. Vallee, J. Cell Biol. **132** (1996) 617.
[6] A.P. Enos and N.R. Morris, Cell **60** (1990) 1019.
[7] V.I. Gelfand and A.D. Bershadsky, Annu. Rev. Cell Biol. **7**, (1991) 93.
[8] D.M. Glover, C. Gonzalez and J.W. Raff, Sci. Am. **268**, (1993) 62.
[9] D.D. Hackney, Nature **377** (1995) 448.
[10] T. Hirano, Trends Biochem. Sci. **20** (1995) 357.
[11] T.E. Holy and S. Leibler, Proc. Natl. Acad. Sci. USA **91** (1994) 5682.
[12] M.A. Hoyt and J.R. Geiser, Annu. Rev. Genet. **30**, (1996) 7.
[13] A.J. Hunt, F. Gittes and J. Howard, Biophys. J. **67**, (1994) 766.
[14] A.A. Hyman and T.J. Mitchison, Nature **351**, (1991) 206.
[15] S. Inoue and H. Sato, J. Gen. Physiol. **50** (1967) Suppl:259.
[16] K.A. Johnson, Annu. Rev. Biophys. Biophys. Chem. **14** (1985) 161.
[17] M. Kirschner and T. Mitchison, Cell **45** (1986) 329.
[18] D. Koshland and A. Strunnikov, Ann. Rev. Cell Dev. Biol. **12**, (1996) 305.
[19] J.R. Lamb, W.A. Michaud, R.S. Sikorski and P.A. Hieter, EMBO J. **13** (1994) 4321.
[20] J. Lechner and J. Carbon, Cell **64** (1991) 717.
[21] R.J. Leslie and J.D. Pickett–Heaps, J. Cell Biol. **96**, (1983) 548.
[22] V.A. Lombillo, C. Nislow, T.J. Yen, V.I. Gelfand and J.R. McIntosh, J. Cell Biol. **128** (1995) 107.
[23] V.A. Lombillo, R.J. Stewart and J.R. McIntosh, Nature **373** (1995) 161.
[24] E.M. Mandelkow, E. Mandelkow and R.A. Milligan, J. Cell Biol. **114** (1991) 977.
[25] J.R. McIntosh, Cold Spring Harb. Symp. Quant. Biol. **56** (1991) 613.
[26] J.R. McIntosh, in: J.S. Hyams and C. Lloyd (eds.), *Modern Cell Biology* (Wiley–Liss, New York, N.Y., 1994) pp. 413.

[27] J.R. McIntosh, P.K. Hepler and D.G. Van Wie, Nature **224** (1969) 659.

[28] J.R. McIntosh and G.E. Hering, Annu. Rev. Cell Biol. **7** (1991) 403.

[29] P.A. McNeil and M.W. Berns, J. Cell Biol. **88** (1981) 543.

[30] P.B. Meluh and M.D. Rose, Cell **60** (1990) 1029.

[31] T.J. Mitchison, J. Cell Biol. **109** (1989) 637.

[32] T.J. Mitchison, Science **261** (1993) 1044.

[33] R.B. Nicklas, Annu. Rev. Biophys. Biophys. Chem. **17** (1988) 431.

[34] R.B. Nicklas, Science **275** (1997) 632.

[35] C. Nislow, V.A. Lombillo, R. Kuriyama and J.R. McIntosh, Nature **359** (1992) 543.

[36] C.M. Pfarr, M. Coue, P.M. Grissom, T.S. Hays, M.E. Porter and J.R. McIntosh, Nature **345** (1990) 263.

[37] A.F. Pluta, A.M. Mackay, A.M. Ainsztein, I.G. Goldberg and W.C. Earnshaw, Science **270** (1995) 1591.

[38] J.W. Raff, Trends Cell Biol. **6** (1996) 248.

[39] C.L. Rieder and E.D. Salmon, J. Cell Biol. **124** (1994) 223.

[40] C.L. Rieder, A. Schultz, R. Cole and G. Sluder, J. Cell Biol. **127** (1994) 1301.

[41] A.D. Rudner and A.W. Murray, Curr. Opin. Cell Biol. **8** (1996) 773.

[42] E.P. Sablin, F.J. Kull, R. Cooke, R.D. Vale and R.J. Fletterick, Nature **380** (1996) 555.

[43] W.S. Saunders, D. Koshland, D. Eshel, I.R. Gibbons and M.A. Hoyt, J. Cell Biol. **128** (1995) 617.

[44] R.V. Skibbens, C.L. Rieder and E.D. Salmon, J. Cell Sci. **108** (1995) 2537.

[45] E.R. Steuer, L. Wordeman, T.A. Schroer and M.P. Sheetz, Nature **345** (1990) 266.

[46] K. Svoboda and S.M. Block, Cell **77** (1994) 773.

[47] K. Svoboda, C.F. Schmidt, B.J. Schnapp and S.M. Block, Nature **365** (1993) 721.

[48] D.A. Thrower, M.A. Jordan, B.T. Schaar, T.J. Yen and L. Wilson, EMBO J. **14** (1995) 918.

[49] E.A. Vaisberg, M.P. Koonce and J.R. McIntosh, J. Cell Biol. **123** (1993) 849.

[50] K.T. Vaughan and R.B. Vallee, J. Cell Biol. **131** (1995) 1507.

[51] I. Vernos, J. Raats, T. Hirano, J. Heasman, E. Karsenti and C. Wylie, Cell **81** (1995) 117.

[52] R.A. Walker and M.P. Sheetz, Annu. Rev. Biochem. **62** (1993) 429.

[53] S.Z. Wang and R. Adler, J. Cell Biol. **128** (1995) 761.

[54] D. Wise, L. Cassimeris, C.L. Rieder, P. Wadsworth and E.D. Salmon, Cell Motil. Cytoskeleton **18** (1991) 131.

[55] L. Wordeman and T.J. Mitchison, J. Cell Biol. **128** (1995) 95.

[56] L.G. Wordeman, E. Steuer, M.P. Sheetz and T.J. Mitchison, J. Cell Biol. **114** (1991) 285.

[57] J.T. Yang, W.M. Saxton and L.S.B. Goldstein, Proc. Natl. Acad. Sci. USA **85** (1988) 1846.

[58] E. Yeh, R.V. Skibbens, J.W. Cheng, E.D. Salmon and K. Bloom, J. Cell Biol. **130** (1995) 687.

[59] T.J. Yen, D.A. Compton, D. Wise, et al., EMBO **10** (1991) 1245.

COURSE 4

THE ROLE OF MICROTUBULES IN THE CREATION OF
ORDER IN THE CELL

Robert L. Margolis

*Institut de Biologie Structurale J.-P. Ebel (CEA/CNRS), 41 avenue des Martyrs,
38027, Grenoble Cedex 1, France*

G. Zaccai, J. Massoulié and F. David, eds.
Les Houches, Session LXV, 1996
De la Cellule au Cerveau
From Cell to Brain: Intra- and Inter-Cellular Communication –
The Central Nervous System

Contents

1. The cytoskeleton and its dynamics

Eukaryotic cells contain cytoskeletal arrays, which are networks of several distinct protein polymers. In the cytoplasm, these polymers fall into three broad classifications; actin filaments, intermediate filaments and microtubules. Actin filaments assemble from monomeric actin protein subunits in an ATP dependent manner [80]. Intermediate filaments are formed by assembly from subunits that belong to one of a family of proteins such as vimentin and cytokeratin [23]. Microtubules assemble from dimeric tubulin protein subunits, hydrolyzing GTP in the process, and form into tubes, the wall of which is comprised of thirteen protofilaments which follow the long axis of the polymer [27]. Each protofilament is a linear array of tubulin subunits. Each of the three major elements of the cytoskeleton is depicted in Fig. 1, as it appears in an interphase mammalian cell.

The different cytoskeletal arrays play roles in generating and organizing diverse structures that are of fundamental importance to the survival of the cell. In general, the cell structure and the arrangement of its organelles derives from the cytoskeleton and its interaction with various intracellular components. However, the cytoskeleton is not necessarily the static structural element that its name implies. Two of the cytoskeletal elements, microtubules and actin fibers, have been shown to frequently engage in highly dynamic interactions with their constituent subunits [27,80]. Less is known about the dynamics of intermediate filaments, although they have also been demonstrated to turnover rapidly in cells [23].

Of these systems, we will consider the role of microtubules, their steady state reactions in the cell, the organelles with which they associate, and the regulatory mechanisms that change their state during the course of the cell cycle.

Many mammalian cell lines have been grown indefinitely in culture after isolation from the host animal. Such cells have served as model systems for the study of the role of microtubules and their involvement in cell control. All cells progress through a sequence of events, known as the cell cycle, during which they duplicate their genome and divide into two daughter cells [70]. Through most of the cell cycle, microtubules do not have a striking appearance. They radiate outward from a central point near the nucleus in an array that is not highly ordered (Fig. 1A). Nor, for the most part, is this interphase array of microtubules necessary to the life of the cell over the short course. Drug treatments that completely block microtubule assembly do not, in general, disrupt progression through the cell cycle from the

Actin network

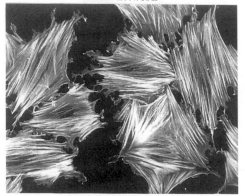

Microtubule network

Vimentin network

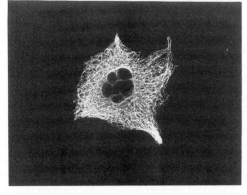

Anti-Tubulin   Propidium Iodide   Merged Image

Fig. 2. The microtubule array and its relationship to the chromosomes in a mitotic cell. The left image shows the microtubule array of a typical mitotic spindle, imaged with anti-tubulin antibody. The spindle is at metaphase, as indicated by the equatorial position of the chromosomes, here visualized with propidium iodide, a DNA specific stain (middle image). The image on the right merges the two different stains obtained on the same cell to show their position relative to each other. All images were obtained by confocal microscopy. The cell used here was BHK, a hamster cell in culture.

time that DNA replication begins until the time the cell reaches mitosis [91]. In the proper genetic background, mammalian cells can reiteratively duplicate their chromatin in the absence of assembled microtubules, and therefore in the absence of effective mitosis [63].

In contrast to the interphase microtubule array, mitotic microtubules are both well ordered and essential to the proper completion of this phase of the cell cycle. Microtubules are the fundamental constituents of the mitotic spindle, which develops from interactions between the two astral arrays of microtubules that emanate from two foci, or poles (Fig. 2). The microtubules from each pole interdigitate where they come in contact with those from the other pole, forming a bipolar organelle that engages the chromosomes, and causes their directed movements. The goal of mitosis is to separate the replicated genome, which has been packaged into individual chromosomes, into two absolutely equal parts, then move them toward the spindle poles where they will be inherited by the two daughter cells that form as a result of the mitotic process.

While the microtubules within the mitotic spindle exhibit a high degree of organization, they at the same time maintain a rapid turnover with their subunits [64]. The organization of the mitotic spindle is the key to its capacity to success-

---

Fig. 1. The different cytoskeletal arrays of an interphase mammalian cell. The actin network in a randomly cycling REF-52 cell line (a rat embryo cell line in culture) displays a cable-like appearance (top image). The microtubule network of the same cell population is less ordered (middle image), but shows highest concentration near the nucleus, the dark circular object in the middle of each cell, where it associates with the centrosome. The intermediate filaments are well extended through the cell's cytoplasm, and, like microtubules, are not highly ordered (bottom image). All images were obtained by immunofluorescence confocal microscopy, using antibodies specific to each of the cytoskeletal elements.

fully separate the genome into two intact and equal sets. Indeed, the survival of the organism depends on the capacity of the mitotic apparatus to complete mitosis with absolute precision. The dynamics displayed by the mitotic spindle therefore cannot be haphazard, but must be part of the design that leads to precise separation of the genome. It can in fact be argued that the dynamics of the mitotic spindle play a fundamental role in creating and maintaining the intricate structure and mechanical activity of the mitotic spindle.

Several drugs bind specifically to tubulin and interfere with microtubule dynamics. Such drugs have recently been shown to block mitotic progression in mammalian cells when used at concentrations sufficient to suppress microtubule dynamics but not sufficient to interfere with the microtubule assembly state [41]. The conclusion of such studies is that a static mitotic spindle can exist, but it cannot function to separate chromosomes. Two major types of polymer dynamics have been described for microtubules. The polymers are capable of "treadmilling", which is to say they exhibit a continuous flux of subunits from one end to the other of the polymer in a steady state reaction [52,53]. They are also capable of rapid length fluctuations at their net assembly end in a process called "dynamic instability" [65].

Both types of microtubule dynamics exist within the mammalian mitotic spindle. In early mitosis, dynamic instability appears to be important to the capture of chromosomes by the mitotic spindle [7,75]. Later in mitosis, microtubules appear to be much less unstable. Instead, they maintain a well ordered array in which a subset of microtubules remains stably associated with the genome at specific binding sites on each chromosome, known as kinetochores. This mature spindle is characteristic of later phases of mitosis where the chromosomes align at the spindle equator (metaphase), or where each chromosome separates into two equal units (chromatids) which move toward the spindle poles (anaphase). During these phases of mitosis, the microtubule network remains very dynamic. But, rather than exhibiting the length fluctuations characteristic of dynamic instability, it displays a constant flux of subunits from the spindle equator toward the two spindle poles [64]. The microtubule dynamics thus behave as a treadmilling system at this phase of mitosis.

2. Origins of microtubule cytoskeletal organization

2.1. Centrosomes/centrioles/basal bodies

2.1.1. The centrosome and microtubule dynamics
Microtubules normally exhibit substantial dynamics in vivo. The typical interphase array of microtubules turns over with a half time of about ten minutes [84].

Since the average cell cycle lasts about 24 hours, this involves complete replacement of the array many times during the life time of the cell. Despite their dynamic nature, microtubules typically radiate outward from a central focus, the centrosome, where one end appears to be tethered [84]. There are several ways that rapid turnover of the polymer array can be consistent with end association at the centrosome.

First, microtubules may associate with the centrosome only for a short time to permit net growth, then may release from the centrosome to permit a net flux of subunits through the polymer. There is evidence that microtubules have this capacity for release, and that released microtubules exhibit substantial treadmilling behavior [76].

Second, microtubules may associate with the centrosome by linkages that make contact with the polymers laterally, leaving the ends free for subunit exchange. Such an arrangement can in principle permit the polymers to flux past a fixed point while undergoing dynamics that are independent of the association, perhaps using a mechanism involving polymer guided diffusion [26]. Does the microtubule actually have the capacity to flux past a lateral association at a fixed point while freely assembling and disassembling? There is strong evidence that microtubule assembly from the kinetochore during mitosis occurs in just this manner. The polymer remains linked to the kinetochore at all times while constantly assembling at this site [45,64]. Thus each subunit that enters the polymer makes its way inexorably from the kinetochore to the centrosome.

Third, microtubules may cycle through rapid phases of net growth and stochastic collapse, in the process of dynamic instability. In this process, each polymer could be imagined as switching between periods of extension or of retraction at infrequent intervals. Such length excursions occur in cells and are most evident in early mitosis where this mechanism may play a role in capture of kinetochores by the microtubule network [7,75].

Since much of microtubule dynamics depends on its interaction with the centrosome, it is important to understand the nature of the centrosome itself. Centrosomes are relatively amorphous membranous elements that sit to one side of the nucleus in interphase cells (Fig. 3A), and act as microtubule organizing centers. Centrosomes are designated as pericentriolar material because they are often but not always centered on a centriole, which is a complex organelle comprised of well ordered microtubule elements. Typically, the centriole consists of a pair of small cylinders of intricate structure that are connected to each other at right angles. Within each cylinder there are nine triplets of microtubules, and each triplet is organized in a spiral array that is offset relative to the radial axis of the centriole [42].

α-Tubulin   Centrosomes

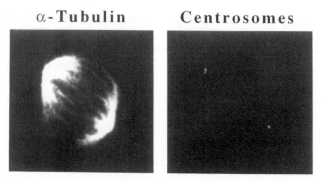

Fig. 3. The position of the centrosomes relative to the microtubules of the mitotic spindle in a mitotic cell. The image at left shows the mitotic spindle with an antibody specific to α-tubulin, one of the two tubulin subunits. The image at right shows the position of the centrosomes, here visualized with an antibody that recognizes a modified form of tubulin (detyrosinated tubulin), which is normally present only at the centrosome. The cell used here was HeLa, a human cervical carcinoma tumor cell in culture. Both images were obtained by confocal microscopy.

2.1.2. The centriole-basal body

The centriole does not template the normal intracellular microtubule array. Indeed, there are many examples of mammalian cells, rodent meiotic cells for example, that lack centrioles at the focus of their centrosomes and yet function normally in assembling microtubule arrays [42]. Instead, templating occurs in the pericentriolar material – an amorphous cloud of proteins with microtubule seeding and controlling functions [29,25].

Despite its lack of immediate involvement in assembly of the microtubule array in the cell, there are circumstances in which the centriole can function to organize microtubule assembly. It is functionally equivalent to, and interchangeable with, the basal body [42]. The basal body has a structure essentially identical to that of the centriole, but functions as the seed to assemble highly organized arrays of microtubules within the cilia and flagellae [28], organelles that protrude from cell surfaces and generate motility by generating oscillatory wave motions. The flagellum of the sperm is a good example of such a structure. The basal body is found at the base of the flagellar microtubules where they exit the cell surface, and is absolutely required for assembly of these microtubule arrays. Flagellar microtubule arrays consist of nine precisely ordered doublets that extend from the outer two elements of the nine triplets in the basal body. Additionally, they contain a central pair of microtubules.

The function of the basal body is to directly template the precisely ordered microtubule arrays of the flagellae, termed 9 + 2 axonemes [87]. The connection between basal bodies and centrioles is made clearer by observations that basal bodies can give rise to centrioles during differentiation [2,14]. In addition, the sperm basal body becomes the oocyte's centriole following fertilization in many species [82]. Further, in mammalian fibroblast cells, the centriole has been observed to template the formation of a cilium next to the nucleus [92]. It is unclear if there is any specific function for this internal "primary" cilium, but it serves to indicate the primary role of the centriole/basal body in organizing these highly ordered microtubule assemblies at various sites. The means by which the basal body/centriole templates precise microtubule arrays is not known, and the precise templating function has not been duplicated in vitro.

2.1.3. The centrosome constituents and their function in microtubule organization
The centriole does not directly template the assembly of the typical microtubule arrays in mammalian cells, but it appears to play a role in determining the number of seeding sites available within the cell for microtubule assembly. The number of sites is not important until mitosis, when it becomes essential to form a two pole mitotic spindle for the proper completion of chromosome separation. Normally there is one connected pair of centrioles in an interphase cell, forming one site of microtubule association. When the cell arrives in mitosis, it must organize two sites of assembly which will serve as foci for the separation of chromosomes into two daughter cells. As the cell cycle progresses, the number of centrioles must therefore be duplicated, but only once. Little is known about the molecular control of the process of centriole duplication in mammalian cells, although the regulation of centriole duplication is under the control of a pathway that involves the tumor suppressor p53 [24]. Each centriole serves as the site where the formation of new "daughter" centrioles occurs, but it does not directly template their formation [44]. In fact, new centrioles can arise from amorphous material rather than in association with preexisting centrioles. Nonetheless, perhaps because they may serve to signal centrosome assembly, preexisting centrioles have been shown necessary to generate new centrioles in mammalian cells in culture [51].

Several proteins are known to have a specific function in the assembly of microtubule arrays that radiate outward from centrosomes. Proteins that function in microtubule assembly during interphase include pericentrin [16] and γ-tubulin [85]. Antibodies directed against either of these proteins will disrupt the capacity of the centromere to sustain microtubule assembly, as assayed either by microinjection of the antibody or by immunedepletion of the protein in an in vitro microtubule assembly system. Their precise roles are not clear, although γ-tubulin is found in multiple phyla and forms seeds with a distinct helical form by complexing with other proteins [86,97,98]. These seeds have been shown to template microtubule assembly in vitro, and are required for assembly of microtubules from the centrosome [68]. It has been shown that γ-tubulin itself binds to the microtubule minus end, and thus is in a position to support assembly [47].

These two proteins, γ-tubulin and pericentrin, appear to play an essential role in microtubule assembly from the centrosome during interphase and may play a role during mitosis as well, although this function is not as clear. Several other proteins seem to be involved specifically in the assembly of the mitotic spindle during mitosis. They include NuMA [61], cytoplasmic dynein with its dynactin regulatory complex [1], and Eg5 [25]. Of these proteins, dynein and Eg5 have presumed microtubule motor function. Eg5 is a plus-ward directed motor [79], while the dynactin complex and dynein are minus end directed [81,83]. Two of these proteins are not accessible to the centrosome during interphase. Eg5 does not associate with microtubules during interphase, but associates strongly with the mitotic spindle. NuMA is sequestered in the nucleus during interphase, and therefore does not interact with microtubules at this time [49]. It is released when the nuclear envelope dissolves on entry into mitosis and is required for the proper assembly and function of the mitotic spindle. NuMA associates with cytoplasmic dynein in a complex, and both components are required for formation of the spindle aster [61]. The opposing motor protein activities of Eg5 and dynein/dynactin are required to create proper association of microtubules in an astral array [25]. It is important to note that none of the mitosis specific components required for proper microtubule assembly from the spindle poles is an integral component of the centrosome. In fact, it has been argued [25] that the particular requirements for microtubule treadmilling in mitosis demand a system of proteins that laterally associate with microtubules rather than end linked seeds such as γ-tubulin. Consistent with this argument, removal of the centrosome from the spindle by micromanipulation does not disturb spindle pole function [69]. Thus, it is possible that the mechanism for seeding microtubule assembly in mitosis is entirely different than it is for interphase.

Not all organisms utilize a centrosomal system for the assembly of microtubules. Yeast for example have a spindle pole body, which has a distinct structure but an equivalent function. The spindle pole body is an organelle buried in the nuclear envelope, which has a trilaminar structure and serves to template a limited number of highly ordered microtubules [10] on entry into mitosis [31]. Regardless of the precise structure used, the microtubule organizing center is normally closely associated with the nuclear envelope during interphase. Microtubule motor proteins, similar to those found in mammalian cells, with both plus-ward and minus-ward activities are required for proper function of the spindle pole in yeast. Plus-ward motors include CIN8 and KIP1 and the opposing minus-end directed motor is KAR3 [78].

It is noteworthy that not all cells use the centrosome as the site from which all microtubules originate. A study of kidney epithelial cells from the dog has shown that there are almost no microtubules associated with the centrosome, but that most are in random orientations relative to the centrosome [9]. Wing epidermal cells of the fruit fly *Drosophila* do not contain centrosomes, and microtubules originate instead from apical plasma membrane surfaces [67]. In cases of ectopic microtubule assembly from defined sites at the epithelial cell plasma membrane, where microtubules are specifically oriented, γ-tubulin appears to be associated with the microtubule site of assembly [60]. In other cases, the centrosome alone has γ-tubulin, and microtubules appear to assemble at the apically associated centrosome, then displace to the plasma membrane [66].

2.2. Centromeres

2.2.1. Microtubule dynamics and the centromere

During mitosis many of the spindle microtubules are attached to structures at both ends. In addition to their linkage to the centrosome (Fig. 3B), many microtubules attach to a centromere at their other end (Fig. 4). This association is critical to the proper function of the mitotic spindle. Attachment must include all genetic elements by the time the cell prepares to exit mitosis or proper segregation of two equal genomes will fail. The site of attachment at each chromosome is the kinetochore, a specific mitotic organelle elaborated by a genetic element that occurs just once in each chromosome, the centromere [74]. The microtubules of the mitotic spindle are all polar with respect to their assembly dynamics. The net disassembly end of the polymer is associated with the centrosomes, while the net assembly ends, where subunits are taken up in a net manner, either overlap at the equator of the spindle or make attachments with the centromeres [58].

Anti-Tubulin　　　　**Anti-Centromere**　　　　　　**Merged Image**

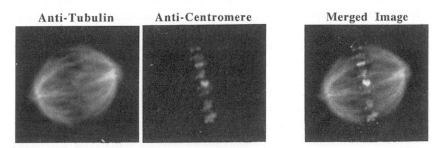

Fig. 4. The position of the centromeres relative to the mitotic spindle at metaphase. As in Fig. 2, the cell is in metaphase. The mitotic spindle is visualized (left image) using anti-tubulin antibody. The central image shows the centromeres, using a human autoimmune antiserum from a patient (GD) with CREST variant scleroderma autoimmune disorder. People with this syndrome characteristically develop antibodies against centromeric proteins, for unknown reasons. The right image is a computer merge of the two left images, to show the juxtaposition of the centromeres relative to the spindle at metaphase. The cell used here was HeLa, a human cervical carcinoma tumor cell in culture. All images were obtained by confocal microscopy.

In mammalia, neither the attachment of microtubules to the centromeres nor association with the centrosomes prevents the continuous flux of the associated microtubules poleward during mitosis through an apparent treadmilling reaction [64]. This capacity to permit constant flow of attached polymers requires a special association of the polymers with each of their attachment sites. The simplest way to understand how polymers may remain associated with specific sites and continue to flow past these sites is to imagine that the polymers associate with these sites through lateral binding molecules which have the capacity to continuously diffuse on the polymers toward their net assembly ends [45]. The molecular detail of such associations is not known, but we have speculated on how such a machinery may operate [26].

The capacity of microtubules to remain attached and to flow at the same time is critical to the proper function of the mitotic spindle. It has been shown that if the mitotic spindle dynamics are blocked by microtubule specific drugs at concentrations that do not affect spindle morphology, the mitotic spindle cannot function to move chromosomes [41]. It will therefore be essential to reach an understanding concerning the molecular detail of this critical process.

2.2.2. Composition of the centromere

The centromere occurs only once per chromosome, because specific DNA sequence is required for the assembly of the proteins required to create the active structure. The genetic composition of the centromeric DNA is not well understood in mammalia, although functional centromeres have recently been constructed from known polynucleotide stretches which include the alpha satellite DNA that is associated with the centromere region of human chromosomes [34]. The alpha satellite DNA sequence is a reiteration of a 171 nucleotide motif. Within this motif are buried sequences that are necessary for centromere assembly. Only one reiterated sequence has a known function. A 17mer DNA consensus motif binds specifically to a well characterized centromeric protein, CENP-B [57].

The centromere associated proteins in mammalian cells include CENP-B, CENP-A and CENP-C [74]. The functions of these proteins are not as yet clear, although CENP-A is a member of the family of histones [73], which are proteins that package DNA into discrete unit structures, the nucleosomes. CENP-B is related to a family of transposons, proteins that are involved in DNA rearrangements [43], and may have primary functions unrelated to maintenance of the centromere. CENP-C does not have obvious homologies among other proteins, but experimental analysis shows that it is absolutely required for the construction of the kinetochore during mitosis [90]. CENP-A and CENP-B are presumably important to centromere function, since proteins that are homologous in primary sequence and in function are also present in yeast centromeres [88,32].

In contrast to mammalia, the molecular composition of the chromatin in centromeres in two different species of yeast has been extensively analysed. In the budding yeast, *Saccharomyces cerevisiae*, the centromeric DNA is comprised of a very short stretch of DNA, of approximately 220 base pairs, which is entirely sufficient to carry out all centromere functions [8,39]. Within this DNA stretch there is a discrete binding site for a complex of several proteins, CBF3 [46]. Among the CBF3 proteins only skp1 has been found to have a homologue in mammalia. These proteins include those that bind to microtubules directly [40], and also support the cell cycle dependent association of a microtubule motor protein [62]. The centromere complex includes at least two cell cycle regulatory proteins. In addition to skp1, there is mad2 [12,33]. Homologues of both mad2 and skp1 are present in mammalia. Skp1 has been shown to complex with a cyclin dependent protein kinase that is required for cell cycle progression [95], and mad2 appears to intervene in centromere dependent regulation of the progression of the cell from metaphase, where the chromosomes are aligned at the equator of the mitotic spindle, to anaphase, where the two sets of chromosomes migrate apart to form the genetic core of the two daughter cells that will form as a result of mitotic completion [48].

The centromere of another yeast, *S. pombe*, is quite different in structure. It contains orders of magnitude more DNA than the *S. cerevisiae* centromere. The DNA is arranged into families of repeat motifs that bracket a core domain [11]. The core domain is absolutely required for centromere function, and probably serves as the site for microtubule association. However, no *S. pombe* centromere associated proteins have yet been described. Of the repeat regions, a small portion must be present on either side of the core for it to function as a centromere [5].

Microtubule motor proteins also associate with the centromeres of mammalian cells, but their functions are not well distinguished. A motor activity appears to be responsible for the rapid migrations of chromosomes between the poles and the equator of the spindle in early mitosis [75]. Further, a protein component of the dynactin microtubule associated motor protein complex, p50, associates with centromeres and is required for proper alignment of chromosomes on the spindle during mitosis [18]. It is, however, more difficult to know if a centromere associated motor protein might have a function in later stages of mitosis when the chromosomes separate.

The description of the microtubule associated proteins of the centromere and of the centrosome, even when it is complete, will not give the entire picture of the means by which microtubules maintain a specific dynamic state during the course of mitosis. There are also microtubule associated proteins such as MAP4 and STOP that attach to microtubules of the mitotic spindle throughout their length, and can serve to regulate microtubule dynamics [71,54]. Additionally, the protein stathmin (also called Op18) does not bind to microtubules but binds with high affinity to the free tubulin subunit, and has been shown to dramatically alter microtubule dynamics in vitro [6]. When expressed in constitutively active form in cells, stathmin disrupts spindle organization [55].

2.3. *Independent and highly ordered microtubule arrays*

Microtubules are also capable of forming elaborate arrays that are apparently independent of centrosome, basal body or centromere control. The prevailing motif for such arrays is that they align in parallel in association with a cell membrane. Such arrays are common in protozoa, where they underlie the cortex in a meshwork that is called a kinetid [50]. They may also form the framework of subcellular structures in protozoa such as the oral apparatus and may have derived from the microtubule array underlying this structure during the course of evolution [20]. For such single celled organisms, structures and specific morphological features must be created by the use of subcellular organelles such as microtubules. The use of microtubules as scaffolds and struts to create and maintain molded structure is commonplace. The use of the kinetid is not much different in principle than the use of a skeletal framework within a dirigible. Such arrays of microtubules following the cell cortex are also common in plant cells, and appear to be important to deposition of cellulose which determines the final shape of the cell [72]. A very specific cortical band of microtubules, the preprophase band, appears before mitosis and disappears as mitosis progresses in plant cells [30]. It serves to position the phragmoplast that will separate the two daughter cells during division.

In a system much closer to mammalia, a frog (*Xenopus laevis*) oocyte contains a cortical array of microtubules that aligns in parallel at the surface of the egg [21]. This parallel array drives rotation of the cortex during the first division cycle probably through association with kinesin, a microtubule motor protein [36]. This rotation is absolutely critical to the correct development of the organism, and delivers a protein important to morphogenesis, catenin, to a specific site [77]. Failure to rotate causes highly aberrant embryogenesis. Although in this case the association with microtubule motors, which serve to link microtubules to membrane bound structures, may contribute to the association with the cortex, nothing is known about the means by which the array comes to align in parallel and in the correct orientation to perform its function.

Other examples of microtubule dependent ordering of intracellular elements abound. For example, the golgi apparatus, the centrally located membranous organelle responsible for packaging products for transport to the cell surface, associates with the centrosome and depends on the microtubules that emanate from the centrosome for its structure [93]. The cell is filled with a membranous array, the endoplasmic reticulum, where proteins are produced by translation of messenger RNA. The endoplasmic reticulum is also ordered in space by the microtubule network [89,13]. Additionally, the mitochondria, the organelles that serve as the power source for the cell, are aligned along microtubules [4]. The nucleus of the sperm is surrounded by a highly ordered array of microtubules, the manchette [59], and analysis of a mouse mutant suggests that this manchette array is required to give the mature shape to the sperm head [15]. All platelet cells and the erythrocytes (red blood cells) of many vertebrates contain an array of microtubules that follow the outer cortex. The cells are disk shaped and the microtubules form a ring that follows the large diameter of the disk. This ring, called the marginal band, appears to stabilize the round flat shape of the cell [22,35].

2.4. Microtubule based communication and ordering of cell space

The capacity of microtubules to associate with intracellular membranes and to align in patterns as exhibited in many systems, most notably protozoa, offers strong evidence for the role of the microtubule cytoskeleton as a participant in the generation of morphology at the single cell level.

During mitosis in mammalian cells, microtubules must accomplish the fundamental task of communicating the position of the mitotic spindle to the cell cortex since cleavage must occur at precisely the position of the equator of the mitotic spindle. This task requires the transfer of a signal from the microtubule network to the cell cortex. This system of microtubule dependent communication and of ordering cell space is both fundamental to cell physiology and amenable to dissection by cell and molecular biology techniques.

Cell cleavage requires deformation of the cell into two equal parts. This event must occur with precise timing and positioning relative to spindle position and progression so that the genome is equally inherited by the two daughter cells that are in the process of formation.

The physical basis of the transfer of message is presently unknown. There is, however, an attractive candidate system for this purpose. A group of proteins known as passenger proteins associate with the centromere during early mitosis, and release from the centromere after anaphase commences [17]. After their release they migrate along the microtubules of the mitotic spindle toward the net assembly end, which is to say toward the zone of overlap between the two astral arrays that comprise the spindle. In mid to late anaphase, these proteins arrive at the equator of the mitotic spindle and then expand outward to form a disk structure, designated the telophase disc, that makes direct contact between the mitotic spindle and the cell cortex [3]. Cell cleavage ensues by recruitment to the telophase disc of actin, myosin and other proteins required to generate the contraction necessary to constrict the cell at the position of the spindle equator (Fig. 5) [56].

Evidence that the chromosomes originate the signal required for cell cleavage in mammalian cells has been obtained from experiments in which the number of

A c t i n   **α-TD-60**

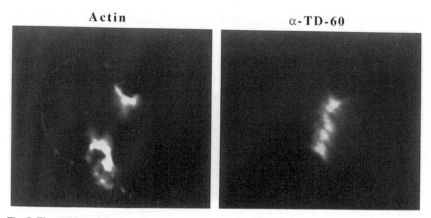

Fig. 5. The position of the ring of actin relative to the telophase disc in a cleaving cell. The image at left shows invaginations on either side of a cell that is in the process of cleavage. Actin, here visualized with fluorescent phalloidin, a compound that binds specifically to assembled actin, is positioned at the site of invagination. This image is a two dimensional slice of the cell. In three dimensions, the actin array would be evident as a ring at the site of cleavage. The image at right shows the position of the telophase disc relative to the actin array in the same cleaving cell. The telophase disc, here visualized with a human autoimmune antiserum that recognizes the TD-60 component protein, appears as a line at the position of the mitotic spindle equator at late mitosis. The two termini of the line are in contact with the actin ring array. Again, this is a two dimensional slice of the cell. In three dimensions the line of TD-60 would be evident as a disc in contact with the ring of actin on its outer edge. The cell used here was HeLa, a human cervical carcinoma tumor cell in culture. Both images were obtained by confocal microscopy.

spindle poles has been increased by micromanipulation [94,19]. It is clear from these experiments that a chromosome must be present between two spindle poles for them to generate a cleavage signal at the position equidistant between the two poles. It is also clear that not all eukaryotic cells use the same signalling mechanism, with the signal originating at the centromeres. Recently, insect spermatocytes have been shown to generate a signal for cleavage whether chromosomes are present or absent in the spindle [96]. In this case, it is evident that the signal must originate elsewhere, perhaps at the spindle poles, and migrate on the mitotic spindle during anaphase. The common element may be that a signal generated on the mitotic spindle might collect at the spindle equator and induce the machinery for cleavage to assemble at the appropriate time.

Microtubules are probably involved in the generation of a variety of cell shape changes, and for this must interact with membranes and other elements of the cytoskeleton. For the well organized mitotic spindle, it is evident that orientation of other structures relative to the position of the centrosomes and of the centromeres can be critical to organization both in time and space. For the less well organized interphase arrays of microtubules, it is perhaps less evident how microtubules can contribute to cell organization. However, the single centrosome creates a natural symmetry breaking element in the interphase cell, and can perhaps play a role in the development of cell polarity. This is perhaps clearest in the case of some epithelial cells that require a specific placement of the centrosome and associated microtubules in order to form a spatially differentiated apical end [66]. The centrosome can also be of importance during development since it can serve as the unique site of localization of elements that must be inherited by only some progeny cells. Localization of developmental cues has been studied in the flat worm *C. elegans*. The centrosome must rotate 90° during the mitotic cycle in the early embryo for the cleavage plane to be in position to distribute the developmental elements to only one daughter cell. This rotation depends on the microtubule aster that emanates from the centrosome [38]. Similar displacements of the microtubule network and of the centrosomes accompany polarization of cells in the early mouse embryo [37].

In summary, the microtubule cytoskeleton is critical to the development of organization, both at the intracellular level and at the level of the organism. Microtubules are in turn organized by association with specific organelles that control their assembly state and dynamics. Microtubules can also form highly ordered arrays independent of these organelles, and usually do so in lateral association with intracellular membranes. The means by which microtubules are ordered in space by their controlling organelles is becoming clearer. The means by which microtubules in turn create order and asymmetry has not been worked out except in the barest outline, and this is the subject of much current research.

Acknowledgements

The author wishes to thank Paul Andreassen, Marc Trielli and Stephanie Martineau for their valuable assistance in assembling the figures used in this manuscript.

References

[1] V. Allen, Curr. Biol. **6** (1996) 630.
[2] K.G.W. Anderson and R.M. Brenner, J. Cell. Biol. **50** (1971) 10.
[3] P.R. Andreassen, D.K. Palmer, M.H. Wener and R.L. Margolis, J. Cell Sci. **99** (1991) 523.
[4] E.H. Ball and S.J. Singer, Proc. Natl. Acad. Sci. USA **79** (1982) 123.
[5] M. Baum, V.K. Ngan and L. Clarke, Mol. Biol. Cell **5** (1994) 747.
[6] L.D. Belmont and T.J. Mitchison, Cell **84** (1996) 623.
[7] L.D. Belmont, A.A. Hyman, K.E. Sawin and T.J. Mitchison, Cell **62** (1990) 579.

[8] K.S. Bloom and J. Carbon, Cell **29** (1982) 305.
[9] M.H. Bre, T.E. Kreis and E. Karsenti, J. Cell Biol. **105** (1987) 1283.
[10] B. Byers, K. Shriver and L. Goetsch, J. Cell Sci. **30** (1978) 331.
[11] C. Clarke and M.P. Baum, Mol. Cell Biol. **10** (1990) 1863.
[12] C. Connelly and P. Hieter, Cell **86** (1990) 275.
[13] S.L. Dabora and M.P. Sheetz, Cell **54** (1988) 27.
[14] E.R. Dirksen, J. Cell Biol. **51** (1971) 286.
[15] G.B. Docher and D. Bennett, J. Embryol. Exp. Morphol. **32** (1974) 749.
[16] S.J. Doxsey, P. Stein, L. Evans, P.D. Calarco and M. Kirschner, Cell **76** (1994) 639.
[17] W.C. Earnshaw and R.L. Bernat, Chromosoma **100** (1991) 139.
[18] C.J. Echeverri, B.M. Paschal, K.T. Vaughan and R.B. Vallee, J. Cell Biol. **132** (1996) 617.
[19] D.M. Eckley, A.M. Ainsztein, A.M. Mackay, I.G. Goldberg and W.C. Earnshaw, J. Cell Biol. **136** (1997) 1169.
[20] K. Eisler, Biosystems **26** (1992) 239.
[21] R.P. Elinson and B. Rowning, Dev. Biol. **128** (1988) 185.
[22] D.W. Fawcett and F. Witebsky, Z. Zellf. **62** (1964) 785.
[23] E. Fuchs and K. Weber, Annu. Rev. Biochem. **63** (1994) 345.
[24] K. Fukasawa, T. Choi, R. Kuriyama, S. Rulong and G.F. Vande Woude, Science **271** (1996) 1744.
[25] T. Gaglio, A. Saredi, J.B. Bingham, M.B. Hasbani, S.R. Gill, T.A. Schroer and D.A. Compton, J. Cell Biol. **135** (1996) 399.
[26] J.R. Garel, D. Job and R.L. Margolis, Proc. Natl. Acad. Sci. USA **84** (1987) 3599.
[27] V.I. Gelfand and A.D. Bershadsky, Annu. Rev. Cell Biol. **7** (1991) 93.
[28] I.R. Gibbons, J. Cell Biol. **91** (1981) 107s.
[29] R.R. Gould and G.G. Borisy, J. Cell Biol. **73** (1977) 601.
[30] B.E.S. Gunning, in: C.W. Lloyd (ed.), *The Cytoskeleton in Plant Growth and Development* (Academic Press, NY) pp. 229.
[31] I.M. Hagan and J.S. Hyams, J. Cell Sci. **89** (1988) 343.
[32] D. Halverson, M. Baum, J. Stryker, J. Carbon and L. Clarke, J. Cell Biol. **136** (1988) 487.
[33] K.G. Hardwick and A.W. Murray, J. Cell Biol. **131** (1995) 709.
[34] J.J. Harrington, G. Van Bokkelen, R.W. Mays, K. Gustashaw and H.F. Willard, Nature Genetics **15** (1995) 345.
[35] G.B. Hayden and A. Taylor, J. Cell Biol. **26** (1965) 673.
[36] E. Houliston and R.P. Elinson, J. Cell Biol. **114** (1991) 1017.
[37] E. Houliston, S.J. Pickering and B. Maro, J. Cell Biol. **104** (1987) 1299.
[38] A.A. Hyman and J.G. White, J. Cell Biol. **105** (1987) 2123.
[39] A.A. Hyman and P.K. Sorger, Annu. Rev. Cell Dev. Biol. **11** (1995) 471.
[40] W. Jiang, K. Middleton, H.J. Yoon, C. Fouquet and J. Carbon, Mol. Cell Biol. **13** (1995) 4884.
[41] M.A. Jordan, D. Thrower and L. Wilson, J. Cell Sci. **102** (1992) 401.
[42] D.R. Kellogg, M. Moritz and B.M. Alberts, Annu. Rev. Biochem. **63** (1994) 639.
[43] D. Kipling and P.E. Warburton, Trends in Genetics **13** (1997) 141.
[44] R.S. Kochanski and G.G. Borisy, J. Cell Biol. **110** (1997) 1599.
[45] D.E. Koshland, T.J. Mitchison and M.W. Kirschner, Nature **331** (1997) 499.
[46] J. Lechner and J. Carbon, Cell **64** (1991) 717.
[47] Q. Li and H.C. Joshi, J. Cell Biol. **131** (1995) 207.
[48] Y. Li and R. Benezra, Science **274** (1995) 246.
[49] B.K. Lydersen and D.E. Pettijohn, Cell **22** (1980) 489.
[50] D.H. Lynn, Biosystems **21** (1988) 299.
[51] A. Maniotis and M. Schliwa, Cell **67** (1991) 495.

[52] R.L. Margolis and L. Wilson, Cell **13** (1978) 1.
[53] R.L. Margolis and L. Wilson, Nature **293** (1981) 705.
[54] R.L.Margolis, C.T. Rauch, F. Pirollet and D. Job, EMBO J. **9** (1990) 4095.
[55] U. Marklund, N. Larsson, H.M. Gradin, G. Brattsand and M. Gullberg, EMBO J. **15** (19) (1996) 5290.
[56] S.N. Martineau, P.R. Andreassen and R.L. Margolis, J. Cell Biol. **131** (1996) 191.
[57] H. Masumoto, H. Masukata, Y. Muro, N. Nozaki and T. Okazaki, J. Cell Biol. **109** (1989) 1963.
[58] J.R. McIntosh and U. Euteneuer, J. Cell Biol. **98** (1984) 525.
[59] J.R. McIntosh and K.R. Porter, J. Cell Biol. **35** (1967) 153.
[60] T. Meads and T.A. Schroer, Cell Motil. Cytoskeleton **32** (1995) 273.
[61] A. Merdes, K. Ramyar, J.D. Vechio and D.W. Cleveland, Cell **87** (1996) 447.
[62] K. Middleton and J. Carbon, Proc. Natl. Acad. Sci. USA **91** (1994) 7212.
[63] A.J. Minn, L.H. Boise and C.B. Thompson, Genes Dev. **10** (1996) 2621.
[64] T. Mitchison, J. Cell Biol. **109** (1989) 637.
[65] T.J. Mitchison and M.W. Kirschner, Nature **312** (1984) 232.
[66] M.M. Mogensen, J.B. Mackie, S.J. Doxsey, T. Stearns and J.B. Tucker, Cell Motil. Cytoskeleton **36** (1997) 276.
[67] M.M. Mogensen, J.B. Tucker and H. Stebbings, J. Cell Biol. **108** (1989) 1445.
[68] M. Moritz, M.B. Braunfeld, J.W. Sedat, B. Alberts and D.A. Agard, Nature **378** (1995) 637.
[69] R.B. Nicklas, S.C. Ward and G.J. Gorbsky, J. Cell Sci. **94** (1989) 415.
[70] C. Norbury and P. Nurse, Annu. Rev. Biochem. **61** (1992) 441.
[71] K. Ookata, S. Hisanaga, J.C. Bulinski, H. Murofushi, H. Aizawa, T.J. Itoh, H. Hotani, E. Okumura, K. Tachibana and T. Kishimoto, J. Cell Biol. **128** (1995) 849.
[72] B.A. Palevitz and P.K. Hepler, Planta **132** (1976) 71.
[73] D.K. Palmer, K. O'Day, H. LeTrong, H. Charbonneau and R.L. Margolis, Proc. Natl. Acad. Sci. USA **88** (1991) 3734.
[74] A.F. Pluta, A.M. Mackay, A.M. Ainsztein, I.G. Goldberg and W.C. Earnshaw, Science **270** (1995) 1591.
[75] C.L. Rieder and S.P. Alexander, J. Cell Biol. **110** (1990) 81.
[76] V.I. Rodionov and G.G. Borisy, Science **275** (1996) 215.
[77] B.A. Rowning, J. Wells, M. Wu, J.C. Gerhart, R.T. Moon and C.A. Larabell, Proc. Natl. Acad. Sci. USA **94** (1997) 1224.
[78] W.S. Saunders, D. Koshland, D. Eshe, I.R. Gibbons and M.A. Hoyt, J. Cell Biol. **128** (1995) 617.
[79] D.E. Sawin, M. LeGuellec, M. Philippe and T.J. Mitchison, Nature **359** (1992) 540.
[80] D.A. Schafer and J.A. Cooper, Annu. Rev. Cell Dev. Biol. **11** (1995) 497.
[81] D.A. Schafer, S.R. Gill, J.A. Cooper, J.E. Heuser and T.A. Schroer, J. Cell Biol. **126** (1994) 403.
[82] G. Schatten, Dev. Biol. **165** (1994) 299.
[83] T.A. Schroer, Curr. Opinion Cell Biol. **6** (1994) 69.
[84] E. Schulze and M. Kirschner, J. Cell Biol. **102** (1986) 1020.
[85] T. Stearns and M. Kirschner, Cell **76** (1994) 623.
[86] T. Stearns, L. Evans and M. Kirschner, Cell **65** (1991) 825.
[87] R.E. Stephens, BioEssays **17** (1995) 331.
[88] S. Stoler, K.C. Keith, K.E. Curnick and M. Fitzgerald-Hayes, Genes Dev. **9** (1995) 573.
[89] M. Terasaki, L.B. Chen and K. Fujiwara, J. Cell Biol. **103** (1986) 1557.
[90] J. Tomkiel, C.A. Cooke, H. Saitoh, R.L. Bernat and W.C. Earnshaw, J. Cell Biol. **125** (1994) 531.
[91] M.C. Trielli, P.R. Andreassen, F.B. LaCroix and R.L. Margolis, J. Cell Biol. **135** (1994) 689.
[92] R.W. Tucker, A.B. Pardee and K. Fujiwara, Cell **17** (1979) 527.

R.L. Margolis

[93] J. Wehland, M. Henkart, R. Klausner and I.V. Sandoval, Proc. Natl. Acad. Sci. USA **80** (1983) 4286.
[94] S.P. Wheatley and Y. Wang, J. Cell Biol. **135** (1996) 981.
[95] H. Zhang, R. Kobayashi, K. Galaktionov and D. Beach, Cell **82** (1995) 915.
[96] D. Zhang and R.B. Nicklas, Nature **382** (1996) 466.
[97] Y. Zheng, M.K. Jung and B.R. Oakley, Cell **65** (1991) 817.
[98] Y. Zheng, M.L. Wong, B. Alberts and T. Mitchison, Nature **378** (1995) 578.

SECTION II. INTRACELLULAR COMMUNICATION
Membranes – Synapses – Time

COURSE 5

INTRACELLULAR MEMBRANE TRAFFIC

Mary McCaffrey [a] and Bruno Goud [b]

[a] *Biochemistry Department, University College Cork, Cork, Ireland*
[b] *Unite Mixte de Recherche CNRS 144, Institut Curie, 26 rue d'Ulm,
75248 Paris Cedex 05, France*

G. Zaccai, J. Massoulié and F. David, eds.
Les Houches, Session LXV, 1996
De la Cellule au Cerveau
From Cell to Brain: Intra- and Inter-Cellular Communication –
The Central Nervous System

Contents

1. Introduction

During the 1960–1970s, an assembly of experimental data established that the eukaryotic cell is formed of an elaborate system of compartments and organelles, each of which possesses a particular composition of proteins and lipids ensuring functional specificity. Thanks to the development of subcellular fractionation and electron microscopy techniques the principal pathways of communication between organelles (secretory/exocytic and endocytic) were characterized. Towards the end of the 1970s, and beginning of the 1980s, the emergence of new approaches – genetic (isolation of temperature sensitive secretory mutants in *Saccharomyces cerevisiae*) and biochemical (*in vitro* reconstitution of inter-organelle transport) – permitted considerable advances in the understanding of the molecular mechanisms involved in the movement of macromolecules between two intracellular compartments. Currently, it is thought that vectorial transport of macromolecules between two compartments, and homeostasis of these compartments, depends on a bidirectional flux of vesicles (Fig. 1). From each organelle, specific mechanisms allow selective packaging into vesicles of proteins and lipids ("cargo") which are destined for another compartment. These vesicles can then specifically recognize the membrane of the acceptor compartment. Having reached these compartments, transported proteins and lipids can either stay there, be transported to another compartment, or return to the donor compartment, if the protein is a macromolecule which normally resides there.

2. Coated vesicles

The model which currently prevails for intracellular membrane traffic is that lumenal proteins, and membrane proteins destined to leave a compartment, are incorporated in vesicles via an interaction, direct or indirect, with cytoplasmic proteins which assemble on the membrane of the donor organelle to form a coat. Coat assembly promotes a change in shape of the donor membrane compartment and induces the formation of transport vesicles (Fig. 1).

Several coat proteins have now been identified. The best characterized are clathrin and the adaptors, implicated in receptor-mediated endocytosis and transport of certain proteins between the Trans Golgi Network (TGN) and endosomes,

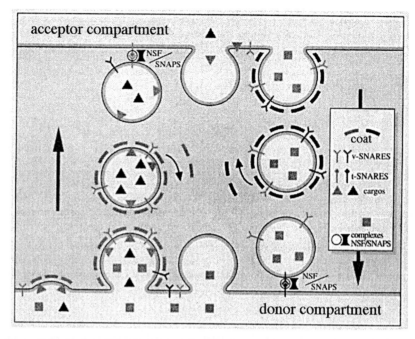

Fig. 1. Inter-organelle transport: Transport between two organelles occurs by the intermediary of a bidirectional flux of vesicles. The first step is vesicle budding resulting from association of cytoplasmic COP proteins with the donor organelle thus forming a coated membrane. Interactions probably occur between macromolecules to be transported (cargo) and the coat proteins. The vesicle then moves towards the acceptor organelle, perhaps guided by the cytoskeleton, and loses its coat, probably before it reaches the target organelle. Association of the vesicle with the target membrane depends on specific interactions occurring between proteins of both donor and acceptor membranes. Cytoplasmic proteins then associate at the interface between the vesicle and the target membrane. The last stage is the fusion of the two membranes, allowing the passage of cargo molecules into the acceptor organelle. Bidirectional flux of vesicles permits the conservation of the lipid and protein composition of donor and acceptor organelles. NSF: N-ethylmaleimide sensitive factor; SNAP: soluble NSF associated protein; SNARE: soluble NSF associated receptor.

and COP I and COP II complexes (for COat Proteins), implicated in the early stages of the secretory pathway. Despite coat diversity, certain remarkable similarities exist between the formation mechanisms of different types of coated vesicles.

2.1. Clathrin coated vesicles

In the case of Clathrin Coated Vesicles (CCVs) forming from plasma membrane, the first stage of their formation requires an interaction between receptors (before or after ligand interaction) and adaptors. This interaction requires specific signals present in the cytoplasmic portion of the receptors (for example, a tyrosine motif). Clathrin triskelions then progressively attach to the adaptor proteins and a series of successive contact rearrangements between clathrin and adaptor molecules ensues (involving other proteins, in particular chaperones) thus causing the invagination of the plasma membrane [1]. Formation of the vesicle is controlled by a GTPase, dynamin, whose function was discovered by studying the *Drosophila* temperature-sensitive mutant called *shibire* (neuronal cells of these mutants accumulate in wells which are covered and do not form endocytic vesicles) [2]. Proteins transported in CCVs developing from the trans-Golgi network, like for example, newly synthesized mannose-6P-receptor associated lysosomal enzymes or molecules of the Major Histocompatibility Class (MHC) II also interact with adaptors, distinct from but structurally similar to the plasma membrane adaptors, via specific signals (tyrosine or di-leucine motifs) [3]. However, dynamin has not been implicated in the formation of these vesicles. On the other hand, a GTPase of the ARF family (ADP-ribosylation factor) controls the process of recruitment of Golgi adaptors and the budding of these vesicles, as in the case of COP I and COP II (see below).

2.2. COP vesicles

Formation of COP I and COP II vesicles has been intensively studied in recent years, in particular by J. Rothman and R. Schekman [4,5]. COP coatomers are formed from a cytosolic complex composed of seven proteins and a protein of the ARF family, ARF1. Recruiting of cytosolic complexes of COP I on membranes requires the activation of ARF1 (exchange GDP-GTP and fixing of ARF-GTP on membranes). Pure ARF1 and COP complexes are necessary and sufficient for the formation in vitro of vesicles from purified Golgi membranes [6]. COP II vesicles, identified until now only in the yeast *S. cerevisiae*, where they have been implicated in transport between the endoplasmic reticulum and Golgi apparatus, are formed by the GTPase Sar1 (displays strong homology with ARF1) and two heterotrimeric complexes, produced from four SEC genes. The sequence of events driving the formation of COP II vesicles is similar to that described for COPI vesicles: recruiting of Sar1-GTP on the endoplasmic reticulum membrane via an interaction with the membrane protein Sec12 which catalyses GDP-GTP exchange; recruiting by Sar1-GTP of Sec 23–Sec24 and Sec 13–Sec31 complexes on the developing vesicles [5].

 Contrary to clathrin coated vesicles, the selective packaging of proteins transported by COP coated vesicles is controversial. Until recently, the predominant idea was that the majority of proteins, utilizing the biosynthetic/secretory pathway, moved rapidly from one organelle to another without an apparent signal (the *bulk flow* hypothesis), and that no relation existed between cargo and the machinery responsible for vesicle formation. The only molecules which escaped from bulk flow were proteins possessing a retention or retrieval signal, like for example resident proteins of the endoplasmic reticulum (signal Lys-Asp-Glu-Leu). However, several experiments contradict the bulk flow hypothesis. For example, one can cite the observation that certain proteins leaving the endoplasmic reticulum are apparently concentrated near sites of formation of COP I vesicles [7]. Nonetheless, deletion of the cytoplasmic domain of several proteins retards considerably their transport. These experiments suggest the existence, as in the case of clathrin coated vesicles, of specific signals present in cargo proteins permitting their interaction with the COP I and the COP II vesicular coats. A direct interaction has already been demonstrated between the Lys-Lys-X-X motifs present on certain transmembrane proteins of the endoplasmic reticulum and the COP I complexes [8].

3. The SNARE hypothesis

Having budded from the donor organelle these vesicles rapidly lose their coat (due to the action of an uncoating ATPase in the case of clathrin coated vesicles forming from the plasma membrane or after GTP hydrolysis associated with ARF for COP I vesicles) and move towards the target organelle. This movement may be guided by microtubules or by the actin cytoskeleton. A series of experiments based on study of the interaction between synaptic vesicles and the presynaptic membrane allowed, in 1993, elucidation, in part at least, of the molecular basis of specific recognition between the vesicle and its target membrane. According to the model called v/t SNARE hypothesis, each vesicle developing from an organelle possesses, on its membrane, a v-SNARE protein (v for vesicle and SNARE for soluble NSF associated receptor), capable of interacting specifically with a t-SNARE (t for target) which is present on the acceptor compartment membrane [4]. Several v/t SNARE "pair" molecules implicated in different stages of membrane transport have now been identified. The v- and t-SNARE molecules are anchored in the membrane via carboxy terminal hydrophobic regions. v-SNARE's are similar to synaptobrevin, a synaptic vesicle protein and t-SNARE's to syntaxin, a protein of the presynaptic membrane. Several laboratories are currently studying mechanisms of interaction between v-SNARE's and t-SNAREs. This interaction comes about notably by the interaction of coiled-coil domains.

The interaction between a v-SNARE and a t-SNARE is followed by the association of several cytosolic proteins allowing the formation of a macromolecular complex at the junction between the vesicle and its target membrane. These proteins include α and γ SNAP (soluble NSF associated proteins) and NSF. NSF is an ATPase. ATP hydrolysis promotes the dissociation of the α/γ SNAP and

NSF complex, preceding a fusion event between the vesicle and target membrane. However, the process of fusion itself remains very badly understood (involvement of fusogenic proteins, role of lipids, . . .?).

4. The Rab family of GTPases

Other proteins play an important role in the mechanisms of anchoring/fusion of vesicles, but their exact function remains unknown. The proteins in question belong to the Rab family of GTPases which belong to the ras superfamily of proteins [9]. We now know of the existence of more than 30 of these proteins which are remarkably conserved in all eukaryotic cells. They are present on the cytoplasmic surface of all organelles, with the exception perhaps of lysosomes, and are anchored in the lipid bilayer by a carboxy-terminal geranyl–geranyl group (a 20 carbon unsaturated chain). Rab proteins circulate between the cytosol and the membrane of vesicles or organelles, the inactive form (GDP-bound) being cytosolic, and the active form (GTP-bound) is essentially membrane bound. The introduction of mutant forms of these Rab proteins (mutations corresponding to oncogenic mutations of ras which block the ras proteins in a GTP-associated or GDP-associated form) profoundly alters cellular events of transport between two organelles and modifies their morphology. Rab proteins, via intermediary effectors which have not yet been identified could perhaps modulate the interaction between v- and t-SNAREs.

5. Conclusion

Despite remarkable progress, accomplished in recent years, several points remain to be elucidated before arriving at a comprehensive view of the molecular mechanisms which regulate the transport of macromolecules between two organelles. Without trying to be exhaustive, the axes of research indicated below promise future development: (1) Identification and characterization of new "coats". COP I coats, the exact *in vivo* function of which is still unknown in biosynthesis/secretory transport (anterograde, retrograde or both?) seems also to be involved in endosomal membrane transport. Others coats involved, for example, in the transport between the Golgi and the plasma membrane still remain to be characterized. (2) Determination of the role of lipids in membrane transport. A growing number of studies indicate a fundamental role of lipids, in particular phosphatidylinositol and its phosphorylated derivatives (IP3, IP4, . . .), in membrane traffic. It is worth mentioning, in this regard, the existence of an essential gene for membrane traffic in yeast (VPS 34) encodes a phosphatidyl inositol-3 kinase, and also the recent discovery that inositol phosphates associate with proteins of the clathrin coat and of COP I. One possibility, among many others, is that the local production of certain lipids could promote formation of vesicles. In this instance, it is necessary to note experiments which show that ARF can activate phospholipase D, an enzyme implicated in the generation of several phosphatidyninositol derivatives [10]. (3) Verification of the v/t-SNARE hypothesis. Despite the fact that this model provides an attractive conceptual frame to explain the specificity of recognition between a vesicle and its target, many uncertainties exist. Other proteins, which remain unidentified, interact in the "bridging" between a vesicle and the target membrane, like that which has been described, for example, in the case of synaptic vesicles and in most stages of membrane transport. The biochemistry of SNARE proteins themselves (transport, recycling, topology, . . .) remains badly understood.

More generally, relations existing between membrane traffic and mechanisms of proliferation or the cell cycle and interactions with the cytoskeleton remain to established. The major challenge of cell biologists in the area of intracellular membrane transport will be, without doubt, the integration of these processes to clarify the molecular basis of inter-organelle communication in normal and abnormal cells.

References

[1] B. Pearse and M. Robinson, Ann. Rev. Cell Biol. **6** (1990) 151.
[2] T. Kosada and K. Ikeda, J. Cell Biol. **97** (1983) 499.
[3] T. Ludwig, R. Leborgne and B. Hoflack, Trends Cell Biol. **5** (1995) 202.
[4] J. Rothman and L. Orci, Nature **375** (1994) 55.
[5] R. Schekman and L. Orci, Science **271** (1996) 1526.
[6] L. Orci, D. Palmer, M. Ahmerd and J. Rothman, Nature **364** (1993) 732.
[7] W. Balch, J. McCaffrey, H. Plutner and M. Farquhar, Cell **76** (1994) 841.
[8] F. Letourneur, E. Gaynor, S. Hennecke, C. Demolliere, R. Duden, S. Emr, H. Riezman and P. Cosson, Cell **79** (1994) 1199.
[9] B. Goud, Médecine/sciences **8** (1992) 326.
[10] P. De Camilli, S. Emr, P. McPherson and P. Novick, Science **271** (1996) 1533.

COURSE 6

I: SIGNALING IN SENSORY CELLS
II: MOLECULAR STRUCTURE AND
FUNCTION OF ION CHANNELS

U.B. Kaupp

Forschungszentrum Jülich, Institut für Biologische Informationsverarbeitung,
52425 Jülich, Germany

G. Zaccai, J. Massoulié and F. David, eds.
Les Houches, Session LXV, 1996
De la Cellule au Cerveau
From Cell to Brain: Intra- and Inter-Cellular Communication –
The Central Nervous System

Contents

1. Signaling in sensory cells

Sensory cells transduce a chemical or physical stimulus into an electrical signal. They are exquisitely sensitive: photoreceptor cells of the vertebrate retina can register single photons and sensory neurons of the olfactory epithelium respond to single odour molecules. At the same time, sensory cells operate over a wide range of light intensities or odour concentrations, i.e., they can adjust their sensitivity to the ambient stimulus intensity. Transduction of a stimulus into a cellular response proceeds in three stages. First, the respective stimulus is registered by specific membrane receptors in the cell, which in turn become active. Second, a highly amplifying cascade of enzymatic reactions either produces or removes an intracellular chemical messenger. Finally, ion channels in the cell membrane either close or open and thereby the cell becomes electrically excited.

Rod and cone photoreceptor cells transduce the absorption of light into a brief, hyperpolarizing voltage pulse by closing cation channels in the plasma membrane. In recent years there have been many striking advances in understanding the physico-chemical basis of photo-electrical excitation in photoreceptor cells. An intracellular messenger, guanosine $3',5'$-cyclic monophosphate (cGMP), carries the message of the absorption event to cGMP-gated ion channels in the surface membrane of the photoreceptor outer segment. Light decreases the cGMP concentration by activating an enzyme cascade that hydrolytically destroys the intracellular messenger. A negative feedback loop involving Ca^{2+} and a guanylyl cyclase (GC) controls the recovery of the photoreceptor cell from the light response. Light also initiates a sequence of events that changes the sensitivity and response kinetics of photoreceptors. These cellular processes – collectively called adaptation – enable the photoreceptor cell to adjust its sensitivity to the ambient illumination. Although we know much less about adaptation, recent experiments suggest that Ca^{2+} ions play an important role by controlling the formation and degradation of cGMP.

cGMP is continuously synthesized and degraded by an enzymatic cycle that involves several proteins. Similarly, Ca^{2+} homeostasis of the cell is controlled by a cycle which involves Ca^{2+} influx through the cGMP-gated channel and subsequent extrusion of Ca^{2+} from the cell by a Na/Ca, K-exchanger. The cGMP- and the Ca^{2+} cycles communicate with each other through a network of enzymatic reactions. A change in one cycle also causes a change in the other, and vice versa.

109

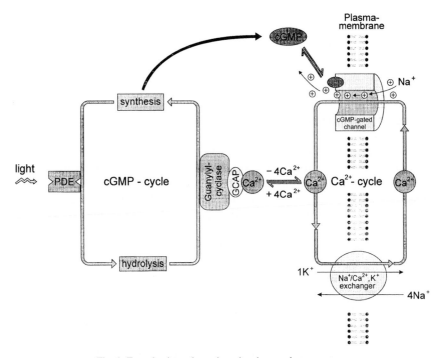

Fig. 1. Transduction scheme in rod and cone photoreceptors.

The gain of the cascade can be as large as $>1,000,000$. For example, one photoactivated rhodopsin molecule can cause the hydrolysis of 10^6 or more cGMP molecules. However, due to the simultaneous onset of inactivating processes, the overall amplification is much lower ($\sim 10^4$–10^5).

These mechanisms endow biological transducers with high reliability, low intrinsic noise, high sensitivity and a large operational range. A brief sketch of the transduction mechanism in rod photoreceptors is shown in Fig. 1.

An interesting variation of the visual transduction scheme exists in sensory neurons of the olfactory epithelium. Odorants bind to specific odorant receptors in the membrane of the chemoreceptive cilia. The receptors in turn activate an adenylate cyclase through an olfactory G-protein, thereby increasing the intracellular cAMP concentration. cAMP binds to cAMP-gated channels in the ciliary membrane. The opening of cAMP-gated channels depolarizes the cell and at the same time Ca^{2+} is rushing into the cell through cAMP-gated channels. Finally, a rise of the intracellular Ca^{2+} concentration activates Ca^{2+}-sensitive Cl^- channels. Thus in contrast to photoreceptor cells, the transduction current in olfactory neurons is carried by two distinct channels, cAMP-gated channels and Ca^{2+}-activated

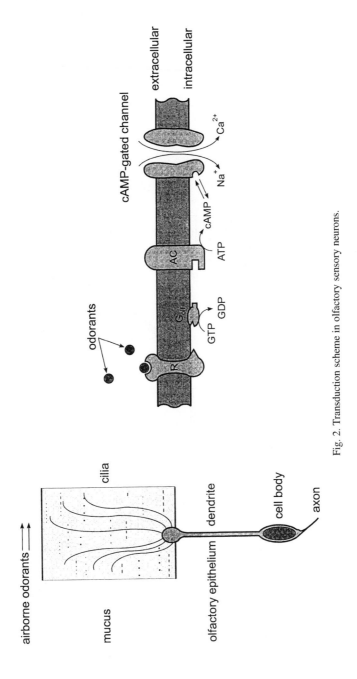

Fig. 2. Transduction scheme in olfactory sensory neurons.

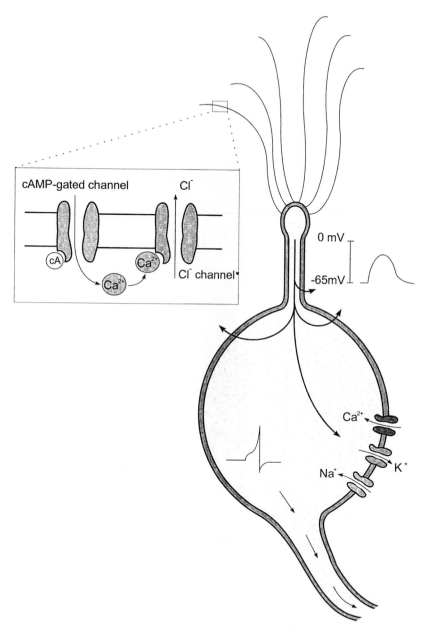

Fig. 3. The transduction current in olfactory cilia is carried by two distinct ion channels, a cAMP-gated, Ca^{2+}-permeable channel and a Ca^{2+}-activated Cl^- channel.

Cl^- channels. A brief sketch of the transduction mechanism in olfactory neurons is shown in Figs. 2 and 3.

References

[1] E.E. Fesenko, S.S. Kolesnikov, and A.L. Lyubarsky, Nature **313** (1985) 310.
[2] J.T. Finn, E.C. Solessio, and K.-W. Yau, Nature **385** (1997) 815.
[3] E.J.M. Helmreich and K.-P. Hofmann, Biochim. Biophys. Acta **1286** (1996) 285.
[4] U.B. Kaupp and K.-W. Koch, Annu. Rev. Physiol. **54** (1996) 153.
[5] K.-W. Koch, Trends Biochem. Sci. **17** (1992) 307.
[6] J.I. Korenbrot, Cell Calcium **18** (1995) 285.
[7] Y. Koutalos and K.-W. Yau, Trends Neurosci. **19** (1996) 73.
[8] D.G. Lambright, J. Sondek, A. Bohm, N.P. Skiba, H.E. Hamm, and P.B. Sigler, Nature **379** (1996) 311.
[9] J.L., Miller, A. Picones, and J.I. Korenbrot, Curr. Opinion Neurobiol. **4** (1994) 488.
[10] A. Polans, W. Baehr, and K. Palczewski, Trends Neurosci. **19** (1996) 547.
[11] E.N. Pugh, Jr. and T.D. Lamb, Vision Res. **30** (1990) 1923.
[12] E.N. Pugh, Jr. and T.D. Lamb, Biochim. Biophys. Acta **1141** (1993) 111.
[13] R.R. Rando, Angew. Chem. **29** (1990) 461.
[14] V.R. Rao and D.D. Oprian, Annu. Rev. Biophys. Biomol. Struct. **25** (1996) 287.
[15] G.F.X. Schertler and P.A. Hargrave, Proc. Natl. Acad. Sci. USA **92** (1995) 11578.
[16] E. Solessio and G.A. Engbretson, Nature **364** (1994) 442.

2. Molecular structure and function of ion channels

Ion channels control the electrical activity of literally every cell. They are particular important for the generation of action potentials in nerve cells and propagation of action potentials along the axons, as well as transmission across synapses. The advent of molecular biology techniques, in conjunction with advances in electrophysiological recording techniques that permit registration of single-channel events in microscopically small patches of the cell membrane, have uncovered a bewildering diversity of structures and mechanisms of activation and modulation of ion channels.

Broadly speaking, ion channels fall into two groups: channels that become activated by a change in membrane voltage (voltage-gated channels), and channels that become activated by binding of a small ligand to a receptor site on the channel polypeptide (ligand-gated channels). The majority of K^+-, Na^+-, and Ca^{2+} channels belong to the superfamily of voltage-gated channels. Curiously, ion channels that are gated by intracellular ligands, i.e. cAMP, cGMP or IP_3, also belong to this large family of channel genes. Many neurotransmitter receptors, for example the nicotinic ACh receptor, the GABA receptor, the glycine receptor, the

5-HT$_3$-receptor, and the glutamate receptor fall into the family of ligand-gated channels.

Functional channels are large heteromeric complexes that are composed of several distinct polypeptides, so-called subunits; the subunits that comprise the complex may be identical or different. Each of these subunits contains functional domains that are responsible for various aspects of channel functioning. Within voltage-gated channels, a positively charged segment in a subunit that resides within the membrane serves as voltage-sensor. Upon depolarization of the voltage, the voltage sensor traverses the membrane and causes a conformational change that gates open the pore of the channel. A short hairpin-like structure lines the interior of the aqueous pore and interacts with the permeating ions. The structure lining the aqueous pore determines which ions can pass through the channel. A ball-and-chain structure serves as an inactivation gate that moves into the mouth of the channel pore to terminate channel activity. The large (180–250 kDa) pore-forming subunits of Na$^+$ and Ca^{2+} channels contain four internal homologous repeats, which fold into a pseudotetrameric structure. K$^+$ channel subunits are roughly 4-fold smaller and assemble into functional tetramers. Each of the repeats of Na$^+$ and Ca^{2+} channels and subunits of K$^+$ channels is characterized by six hydrophobic segments that traverse the membrane. Recently, several subunits have been discovered that form K$^+$-selective pores either by themselves or by co-expression with other K$^+$ channel subunits. These novel subunits have either one or two transmembrane segments. Moreover, subunits have been discovered with either four or eight transmembrane segments and two pore-forming regions.

Within ligand-gated channels, regions in the extra- or intracellular part of the channel protein serve as "receptors" that bind appropriate ligands. Binding of a ligand to a receptor leads to a conformational change within the protein structure that in turn leads to opening of the pore. These receptor domains are targets of many agonists, antagonists, blockers, toxins, peptides, etc. Elucidation of the underlying structures will further our understanding of channel functioning and will aid in the development of selective and sensitive drugs that enhance or suppress ligand-gated channel activity. Many, but not all, subunits forming ligand-gated channels are characterized by four transmembrane segments. In the nicotinic AChR, segment M2 has been identified as pore-lining element. In contrast, glutamate receptor subunits possess only three transmembrane segments and a hairpin-like pore structure between M1 and M2 that becomes inserted into the membrane from the intracellular side. Whereas voltage-gated channels adopt a tetrameric our pseudo-tetrameric structure, ligand-gated ion channels are built from five subunits. For example, the nicotinic AChR is built from four distinct subunits with a stoichiometry of $\alpha_2\beta\gamma\delta$.

References

[1] J.T. Finn, M.E. Grunwald, and K.-W. Yau, Annu. Rev. Physiol. **58** (1996) 395.

[2] B. Hille, *Ionic Channels of Excitable Membranes*, Ed. 2 (Sinauer Associates, Sunderland, 1992).

[3] F. Hofmann, M. Biel, and V. Flockerzi, Annu. Rev. Neurosci. **17** (1994) 399.

[4] M. Hollmann and S. Heinemann, Annu. Rev. Neurosci. **17** (1994) 31.

[5] L.Y. Jan and Y.N. Jan, Nature **371** (1994) 119.

[6] L.Y. Jan and Y.N. Jan, Annu. Rev. Neurosci. **20** (1997) 91.

[7] U.B. Kaupp, Curr. Opinion Neurobiol. **5** (1995) 434.

[8] M. Montal, Annu. Rev. Biophys. Biomol. Struct. **24** (1995) 31.

[9] H.-C. Pape, Annu. Rev. Physiol. **58** (1996) 299.

[10] O. Pongs, Physiol. Rev. **72** (1992) S69.

COURSE 7

THEORETICAL MODELS FOR OSCILLATIONS IN BIOCHEMICAL AND CELLULAR SYSTEMS

Albert Goldbeter

Faculté des Sciences, Université Libre de Bruxelles, Campus Plaine, C.P. 231, B-1050, Brussels, Belgium

G. Zaccai, J. Massoulié and F. David, eds.
Les Houches, Session LXV, 1996
De la Cellule au Cerveau
From Cell to Brain: Intra- and Inter-Cellular Communication –
The Central Nervous System

117

Contents

1. Introduction

Oscillatory phenomena have been observed and studied theoretically in chemical systems [1,2], but appear to be particularly widespread in biological systems [3]. The usefulness of mathematical models in theoretical biology is well illustrated by the study of oscillations in biological systems. The purpose of this paper is to show how different modes of cellular regulation are capable of giving rise to sustained oscillatory behavior. After a brief presentation of biological rhythms and their molecular mechanisms, considered in turn will be examples of oscillations resulting from the regulation of enzymatic activity, receptor function, transport, and gene expression. Theoretical models based on experimental observations will thus be briefly described for glycolytic oscillations in yeast cells, the periodic synthesis of cyclic AMP in *Dictyostelium* amoebae, intracellular calcium oscillations, and circadian rhythms in *Drosophila*. The theoretical approach throws light on the properties shared by these various rhythms characterized by widely different periods.

2. Rhythmic phenomena in biological systems

Rhythms are observed at all levels of biological organization. The most rapid rhythms are those observed in neurons, with periods ranging from 10^{-2} to 10 s [4]. The generation of trains of periodic action potentials is well understood: the phenomenon results from the interactions between several voltage-dependent ionic conductances. Neurons, as well as muscle (among which cardiac) cells, are electrically excitable: an action potential is generated when a suprathreshold depolarizing stimulus causes the membrane potential to abruptly change its polarity before returning to its stable, resting value. In certain conditions, e.g. in the presence of a constant depolarizing current, such an excitability transforms into the generation of repetitive action potentials at regular intervals. These oscillations of the membrane potential appear to play a key role in the functioning of the brain, e.g. in the processing of sensory information. From a theoretical point of view, neural rhythms can be described in terms of equations of the Hodgkin–Huxley type [5], first used to account for the electrical excitability of the squid giant axon in terms of the sodium and potassium conductances.

The cardiac rhythm originates from the autonomous periodic electrical activity of specialized tissues of the heart, such as the sinus or the auriculo-ventricular nodes. At the cellular level, the mechanisms of oscillatory behavior are largely similar to those underlying neuronal oscillations.

Besides these oscillations of electrical origin, other rhythms originate from the various modes of cellular regulation. Thus, several examples of oscillatory enzyme reactions are known, with a period of the order of a few minutes [6,7]. The mitotic oscillator which controls the eukaryotic cell division cycle is an important example of biochemical oscillator involving a cascade of enzymatic reactions regulated by phosphorylation–dephosphorylation, which culminates in the periodic activation of the protein kinase cdc2. The period of the mitotic oscillator varies from several minutes in some embryonic cells up to 24 h or more in somatic cells. Theoretical models for the mitotic control system have been proposed (see Ref. [3] for review).

Oscillations of intracellular calcium of a period ranging from seconds to minutes have been observed since 1985 in a large variety of cell types [8]. Besides oscillations of this intracellular messenger, oscillations of intercellular messengers are known, such as the periodic synthesis of cyclic AMP in *Dictyostelium* amoebae [9], or the pulsatile secretion of a large number of hormones, with periods ranging from some 10 min for insulin, up to 3 h for the growth hormone.

Circadian rhythms, which have a period of about 24 h, are observed in nearly all living organisms, including some bacterial species. Significant experimental advances have been made in recent years as to the molecular mechanisms of these rhythms, particularly in organisms such as *Drosophila* [10].

Other rhythms possess a supracellular mechanism involving regulatory interactions between different organs, as is the case for the ovarian cycle which has a period of about 28 days in the human female, or between different animal species. Predator–prey oscillations represent the first periodic phenomenon that has been studied in a theoretical manner by means of mathematical models, in the first quarter of this century. The study of these oscillations remains a classical problem in theoretical ecology. A particular case is that of epidemics which recur periodically, owing to the interactions between an infectious agent and a population of susceptible hosts which develop an immune response of variable duration against the pathological agent [11,12].

Focusing on biochemical rhythms of a nonelectrical nature, we will examine by means of a few selected examples how oscillations originate from the different modes of cellular regulation, and will underline the role of theoretical models in the study of rhythmic behavior. We shall also mention the conditions in which simple periodic behavior gives rise to more complex oscillatory phenomena, including chaos.

3. Cellular regulation and oscillatory behavior

Cellular regulation is exerted in a variety of ways to control the activity of enzymes, the functioning of receptors and transport processes, and the expression of genes associated with the synthesis of particular proteins. Each of these types of cellular regulation can give rise to rhythmic behavior.

3.1. Enzymatic regulation: glycolytic oscillations

Glycolysis is an important metabolic pathway, the function of which is to synthesize ATP upon degradation of a sugar such as glucose. It is known for four decades that upon addition of glucose, damped oscillations of NADH (a glycolytic intermediate whose fluorescence can be recorded in a continuous manner) occur in a yeast cell suspension. Experiments carried out in yeast extracts later showed that these oscillations become sustained when a glycolytic substrate such as glucose or fructose is injected at a constant rate. Glycolytic oscillations remain the prototype of periodic behavior due to the regulation of enzyme activity [3,5,6].

The period of the phenomenon decreases from about 8 to 3 min when the substrate input rate increases. Moreover, sustained oscillations occur in a window bounded by two critical values of this control parameter: below 20 mM/h and above 160 mM/h, the system evolves towards a stable steady state.

Glycolysis consists of a chain of enzymatic reactions leading from hexokinase to phosphofructokinase (PFK), then to other reactions eventually producing, in yeast, ethanol and CO_2. Very early on, experiments showed that PFK plays a primary role in the generation of glycolytic oscillations (see Refs. [3,6,7] for review). The oscillations indeed disappear as soon as an intermediate following the step catalyzed by PFK (e.g. fructose 1,6-bisphosphate) is used as glycolytic substrate.

What are the particular properties of PFK which enable this enzyme to generate (in conjunction with a source and a sink) metabolic oscillations? Whereas negative feedback processes are by far the most common in enzymatic regulation, PFK is subject to positive feedback regulation by a reaction product, ADP, and by AMP produced from ADP. The PFK reaction is therefore autocatalytic since the rate of the reaction increases as the product concentration rises. In addition to the nonlinearity associated with this positive feedback process, the PFK kinetics possesses a sigmoidal character due to the allosteric nature of the enzyme: PFK contains several subunits which exist in two conformational states differing by the affinity for the substrate and/or the catalytic activity. A phenomenon of cooperativity characterizes the transition from the less active T to the more active R conformation; this transition is either concerted or sequential. Binding of the product to a regulatory site induces the transition from the T to the R state of the enzyme.

The analysis of a mathematical model for a reaction catalyzed by an allosteric enzyme activated by its reaction product allows one to better comprehend the mechanism of glycolytic oscillations generated by PFK [3,7]. The model is described by a system of two ordinary differential equations, the nonlinearity of which results from the positive feedback exerted by the reaction product and from the cooperativity of the enzyme. The two variables considered are the substrate and product concentrations. The main control parameters are the substrate injection rate, v, and parameters linked to the enzyme, such as its concentration, its maximum rate, or the allosteric constant that measures the ratio of enzyme in the T and R states in the absence of ligand.

The study of the dynamic behavior of the model by means of linear stability analysis and phase plane analysis shows that below a critical value of v and above a second, higher critical value of this parameter, the system evolves towards a stable steady state [3]. In agreement with experimental observations, sustained oscillations of the substrate and product concentrations occur in the window bounded by the two critical values of v. In the phase plane where the concentration of substrate is plotted versus that of the product, these oscillations correspond to the evolution towards a closed curve, known as a limit cycle, which surrounds the unstable steady state. Here, the limit cycle is unique and can be reached regardless of initial conditions. This ensures the stability of this type of periodic behavior in regard to fluctuations.

Whereas simple periodic behavior of the limit cycle type is the only one that can be observed in this model, more complex oscillatory phenomena occur as soon as two instability-generating mechanisms interact within the same system [3]. Thus, when two autocatalytic enzymatic reactions are coupled in series, the variety of modes of dynamic behavior in such a three-variable system is greatly increased [13]. Depending on the values of a single control parameter, we can observe simple periodic oscillations, the coexistence between two regimes of stable periodic oscillations of the limit cycle type, and aperiodic oscillations in the form of chaos. The latter oscillations correspond to the evolution towards a strange attractor in the phase space: the oscillatory system remains confined in a given region of the phase space without ever passing again through any point of the trajectory, as would occur in the case of a limit cycle.

The coexistence between two stable limit cycles illustrates well the interest of a theoretical approach to biological rhythms. This phenomenon, referred to as *birhythmicity* [13], was predicted theoretically before being observed experimentally in a system of coupled oscillatory chemical reactions [14]. Birhythmicity is much less frequent than the coexistence between two stable steady states, known as *bistability*, for which numerous experimental examples are known in chemistry and biology.

While three variables at least are needed to obtain chaos, birhythmicity can already be observed in a two-variable system, as in the two-variable model of an autocatalytic enzyme reaction proposed for glycolytic oscillations, when this model is extended to take into account a nonlinear recycling of product into substrate [3].

3.2. Receptor regulation: oscillations of cyclic AMP in Dictyostelium amoebae

The generation of periodic signals of cyclic AMP (cAMP) in the amoebae *Dictyostelium discoideum* represents a model of choice for pulsatile intercellular communication in higher organisms. Following starvation, these amoebae collect around aggregation centers by a chemotactic response to cAMP signals emitted by the centers with a periodicity of about 5 min. On agar, the amoebae aggregate around the centers by forming concentric or spiral waves. Experiments in stirred cell suspensions confirm the periodic nature of cAMP synthesis in this slime mould [9].

The mechanism responsible for the pulsatile generation of cAMP signals again involves a positive feedback loop: extracellular cAMP binds to a cell surface receptor and thereby activates the enzyme adenylate cyclase which catalyzes the synthesis of cAMP from ATP. Intracellular cAMP thus produced is secreted into the extracellular medium where it can bind again to the receptor. This mechanism of self-amplification would lead to a biochemical "explosion" were it not for limiting factors that counteract the effect of autocatalysis. In the case of glycolytic oscillations, substrate consumption plays such a limiting role. In *Dictyostelium*, receptor regulation is the process that limits the self-amplification in cAMP synthesis. As soon as cAMP binds to the active form of the receptor, the latter is phosphorylated. This reversible phosphorylation accompanies the transition of the receptor into a desensitized state unable to elicit the activation of adenylate cyclase. Experiments indicate that cAMP oscillations are accompanied by a periodic alternation of the receptor between the active (nonphosphorylated) and desensitized (phosphorylated) states [15].

The analysis of a mathematical model based on self-amplification in cAMP synthesis and on the reversible desensitization of the cAMP receptor allows one to determine the conditions in which this system of intercellular communication operates in a periodic manner [3,16]. The model, described by a system of three nonlinear kinetic equations, gives rise to sustained oscillations in the concentrations of intra- and extracellular cAMP, and in the fraction of active, nonphosphorylated receptor.

This model not only accounts for the periodic nature of cAMP synthesis but also provides an explanation for the onset of cAMP oscillations in the course of *Dictyostelium* development. Soon after starvation, amoebae are not capable of amplifying cAMP signals; after a few hours, amoebae begin to relay these signals by amplifying them in a pulsatile manner. Still a few hours later, the amoebae acquire the capability of generating pulsatile cAMP signals in a periodic, autonomous manner. The theoretical model shows that these developmental transitions *no re-lay–relay-oscillations* correspond to transitions between different modes of dynamic behavior resulting from the continuous increase in the control parameters measuring the activity of adenylate cyclase and phosphodiesterase, the enzymes that synthesize and degrade cAMP, respectively. The cells would thus follow a *developmental path* in this parameter space; cells most advanced on such a path would be the first to enter the oscillatory domain and would become aggregation centers capable of releasing autonomously periodic signals of cAMP [3]. This explanation bears, more generally, on the ontogenesis of biological rhythms, as it shows how the continuous variation of biochemical parameters or ionic conductances can lead to the passage through a bifurcation point corresponding to the onset of periodic behavior in the course of development.

When including the effect of diffusion of cAMP in the extracellular medium, the system of equations describing the cAMP signaling system includes a partial differential equation for extracellular cAMP. In such conditions, computer simulations show [17] that the desynchronization of cells on the developmental path followed in parameter space after starvation provides a plausible mechanism for the spontaneous formation of spiral waves of cAMP which are observed, along concentric waves, in the course of *D. discoideum* aggregation.

The cAMP signaling system in *Dictyostelium* can also serve as model for intercellular communications of a pulsatile nature. Thus, most hormones are secreted not constantly but rather in a pulsatile manner, with a periodicity ranging from about 10 min up to several hours; such oscillations of relatively high frequency are often superimposed on a slower, circadian variation. The prototype of pulsatile hormone secretion is that of GnRH. This hormone, secreted by the hypothalamus with a periodicity of 1 h in the rhesus monkey and man, induces the secretion by the pituitary of the gonadotropic hormones LH and FSH.

In *Dictyostelium*, the signals of cAMP are encoded in terms of their frequency: while signals emitted every 5 min induce cell aggregation and differentiation, constant signals or cAMP pulses emitted every 2 min fail to have such physiological effects. The model shows that if the receptor has enough time to resensitize when the interval separating two pulses is of 5 min, it cannot resensitize sufficiently when the interval if of 2 min only, or when the signal is applied in a continuous manner [3]. Similarly, the frequency of GnRH secretion governs the physiological efficacy of the hormone: while a GnRH signal emitted once an hour induces the normal levels of LH and FSH required for ovulation, signals of a periodicity of 30 min or 2 h, as well as constant GnRH signals, remain ineffective. These results, which possess important clinical implications, can be accounted for theoretically in terms of a model in which a receptor undergoing reversible desensitization is subjected to a pulsatile signal of variable frequency [18].

3.3. Transport regulation: oscillations of intracellular calcium

Among cellular rhythms discovered in recent years, few are as important and widespread as calcium oscillations [8]. These oscillations were first demonstrated in 1985 in mouse eggs upon fertilisation, and have since been observed in a variety of cell types (hepatocytes, cardiac or pancreatic cells, pituitary gonadotrophs, . . .), following stimulation by a hormone or a neurotransmitter. In view of their ubiquity and of the key role of calcium as intracellular messenger, calcium oscillations represent one of the most significant advances over the last decade in the field of cellular signaling.

The molecular mechanism of the oscillations relies on the transport of calcium from intracellular stores – the endoplasmic reticulum, or the sarcoplasmic reticulum in muscle and cardiac cells – into the cytosol. When a cell is stimulated by a hormone, the binding of this ligand to its receptor elicits the synthesis of an intracellular messenger, inositol 1,4,5-trisphosphate (IP$_3$), which triggers the release of calcium from the intracellular stores; cytosolic calcium is then pumped back into the stores. Hormonal stimulation thus triggers a transient increase in cytosolic calcium concentration.

The fact that the response to hormonal stimulation often takes the form of sustained calcium oscillations is due to the regulation of calcium transport known as *calcium-induced calcium release* (CICR): cytosolic calcium activates the release of calcium from intracellular stores. This regulation also represents an example of receptor regulation, since the IP$_3$ receptor serves as channel allowing the efflux of calcium from the endoplasmic reticulum.

The analysis of theoretical models based on CICR again shows how this nonlinear regulatory process gives rise to an instability of the steady state, associated with the evolution to sustained oscillations, for appropriate values of the control parameters [3,19]. The simplest model based on CICR contains two variables, namely the concentrations of cytosolic and intravesicular calcium. This model accounts for the effect of a progressive increase in extracellular stimulation: below a critical value of the external signal, cytosolic calcium reaches a low steady state level. In a window bounded by two critical values of the external stimulus, oscillations occur with a frequency that increases with the level of stimulation. Above the higher critical stimulus value, cytosolic calcium reaches a high steady state level. A variety of theoretical models based on the detailed regulation of the IP$_3$ receptor and on CICR have recently been proposed.

The study of theoretical models has been extended to the propagation of calcium waves within cells (see, e.g., Ref. [20]). Such waves are observed in numerous cell types such as cardiac myocytes, eggs, hepatocytes, and endothelial cells.

3.4. Genetic regulation: circadian rhythms in Drosophila

Circadian rhythms, which have a period close to 24 h, are encountered in nearly all living organisms, including some bacterial species, and possess an important physiological function in allowing the organism to adapt to its periodically changing environment. In humans, many functions vary in a circadian manner, as illustrated for example by the sleep–wake cycle and by nutrition. In view of the circadian variation of a large number of hormones and enzymatic activities, the response of the organism to some drugs may also vary according to the time of the day; this observation sets the biochemical foundations for the rapidly growing field of chronopharmacology.

Like the other oscillatory phenomena mentioned above, circadian rhythms are endogenous, i.e. they originate from regulatory processes within the organism rather than from the periodic variation of the environment. Circadian rhythms indeed generally persist in constant light or darkness. Among the most conspicuous properties of circadian rhythms are the possibility of entrainment by light–dark or temperature cycles, and the relative independence of their period with respect to temperature, a phenomenon referred to as *temperature compensation*.

In mammals, circadian rhythms are generated by the suprachiasmatic nuclei (SCN) which are groups of neurons located in the hypothalamus. How SCN neurons are able to generate rhythms of 24 h period remains an open question. The phenomenon occurs at the cellular level, since experiments indicate that an isolated SCN neuron retains the circadian variation in electrical activity. The unraveling of the molecular bases of circadian rhythms has undergone rapid advances, thanks to genetic studies, in organisms such as the fly *Drosophila* or the fungus *Neurospora* [10]. A circadian rhythm of locomotor activity has been demonstrated in *Drosophila*. Mutagenesis studies have permitted to isolate "short" and "long" period mutants for which the periodicity of the locomotor rhythm has shifted from close to 24 h in the wild type to 19 h and 28 h, respectively [21]. The mutated gene is known as *per* (for "period") and codes for a protein, PER.

Both the *per* mRNA and the protein PER vary in a circadian manner. The peak in the mRNA precedes by several hours the peak in PER. This observation suggested that the mechanism of oscillations involves a negative feedback exerted by PER on the expression of the *per* gene [22]. Such a mechanism for oscillations in the synthesis of a protein and its mRNA was first proposed by Goodwin [23] on the basis of a theoretical model, shortly after Jacob and Monod laid the molecular foundations of genetic regulation.

Recent studies have shown that the PER protein acts as a regulator of transcription and can thereby influence the expression of a large number of genes. Also involved in the generation of circadian rhythmicity in *Drosophila* is the multiple phosphorylation of PER, which could control the degradation of the protein and/or its entry into the nucleus.

A theoretical model based on the control of PER degradation and entry into the nucleus by two successive phosphorylations and on the negative feedback exerted by PER on the expression of the *per* gene is described by a system of five nonlinear kinetic equations [3,24]. The numerical integration of these equations shows that sustained oscillations in the concentrations of PER and its mRNA can occur for appropriate parameter values. In agreement with experimental observations, the maximum in *per* mRNA precedes the peak in PER by a few hours. The model shows that the successive phosphorylations of the protein introduce a delay in the negative feedback loop; such delays are known to favor the occurrence of sustained oscillations.

Besides PER, a second protein, referred to as TIM, encoded by the gene *tim* (for "timeless"), plays an important role in the mechanism of circadian oscillations in *Drosophila*. The PER and TIM proteins form a complex which migrates from the cytosol into the nucleus where it represses the expression of the *per* and *tim* genes. A theoretical model incorporating the formation of the PER-TIM complex as well as the multiple phosphorylation of the two proteins has been proposed [25]. This extended model, described by a system of ten kinetic equations, yields results similar to those obtained in the simpler model based on regulation by PER alone. However, the domain of oscillations in parameter space is larger. Moreover, besides simple periodic oscillations, the extended model allows for the occurrence of more complex phenomena such as birhythmicity and chaos [25]. The possible physiological significance of these complex modes of oscillatory behavior remains questionable, however, since they occur in restricted ranges of parameter values.

The genetic regulatory mechanism involved in the generation of circadian rhythms in *Drosophila* could apply, more generally, to other organisms. Thus, in *Neurospora*, a similar negative autoregulatory loop has been characterized for the expression of the *frq* ("*frequency*") gene [10]. It is likely that the regulation of gene expression and protein synthesis both form a central part of the mechanism of circadian rhythms for most organisms. Inhibitors of protein synthesis and of DNA transcription indeed induce phase shifts of circadian rhythms, or even their suppression when the addition of such inhibitors exceeds a threshold level.

4. Conclusions

Biochemical as well as neuronal oscillations all result from instability of a nonequilibrium steady state, beyond which the system evolves towards a limit cycle in the phase space. Regulatory processes based on positive or negative feedback lie at the heart of the instability-generating mechanism that underlies the onset of periodic behavior. The interaction between multiple instability-generating mechanisms, often associated with a multiplicity of feedback processes, can lead to more complex oscillatory phenomena including "bursting" oscillations, the coexistence between two (or more) simultaneously stable periodic regimes (birhythmicity), or aperiodic oscillations (chaos).

The study of theoretical models complements well the experimental approach on which these models are based. The interest of such models is to shed light on the core mechanism capable of generating oscillations – by pinpointing the essential variables as well as their interactions – and on the precise conditions in which oscillations occur. Often indeed, verbal descriptions based on sheer intuition alone do not suffice for predicting the dynamic behavior of a complex system containing a large number of variables and governed by multiple, nonlinear regulatory interactions.

Furthermore the theoretical approach underlines the deep unity of biological rhythms. At the cellular level, the latter may originate from the regulation of enzyme activity, receptor function, transport processes, or gene expression. Regardless of the type of control and the nature of the molecules involved, common properties emerge for these rhythms, even if they possess widely different periods.

Acknowledgements

This work was supported by the Programme "Actions de Recherche Concertée" (convention 94/99-180) launched by the French Community of Belgium.

References

[1] G. Nicolis and I. Prigogine, *Self-Organization in Nonequilibrium Systems* (Wiley, New York, 1977).
[2] R.J. Field and M. Burger (eds.), *Oscillations and Traveling Waves in Chemical Systems* (Wiley, New York, 1985).
[3] A. Goldbeter, *Biochemical Oscillations and Cellular Rhythms. The molecular Bases of Periodic and Chaotic Behaviour* (Cambridge University Press, Cambridge, UK, 1996).
[4] J. Jacklet (ed.), *Neuronal and Cellular Oscillators* (Dekker, New York, Basel, 1989).
[5] C. Koch and I. Segev (eds.), *Methods in Neuronal Modeling* (MIT Press, Cambridge, MA, 1989).
[6] B. Hess and A. Boiteux, Annu. Rev. Biochem. **40** (1971) 237.
[7] A. Goldbeter and S.R. Caplan, Annu. Rev. Biophys. Bioeng. **5** (1976) 449.
[8] M.J. Berridge (ed.), *Calcium Waves, Gradients and Oscillations*, CIBA Foundation Symp., (Wiley, Chicester, 1995).
[9] G. Gerisch and U. Wick, Biochem. Biophys. Res. Commun., **65** (1975) 364.

[10] J.C. Dunlap, Annu. Rev. Physiol., **55** (1993) 683.
[11] J.D. Murray, *Mathematical Biology* (Springer, New York, 1989).
[12] R.M. Anderson and R.M. May, *Infectious Diseases of Humans. Dynamics and Control* (Oxford University Press, Oxford, 1992).
[13] O. Decroly and A. Goldbeter, Proc. Natl. Acad. Sci. USA, **79** (1982) 6917.
[14] M. Alamgir and I.R. Epstein, J. Am. Chem. Soc., **105** (1983) 2500.
[15] R.E. Gundersen, R. Johnson, P. Lilly, G. Pitt, M. Pupillo, T. Sun, R. Vaughan, and P.N. Devreotes, in: A. Goldbeter (ed.), *Cell to Cell Signalling: From Experiments to Theoretical Models* (Academic Press, London, 1989) p. 477.
[16] J.L. Martiel and A. Goldbeter, Biophys. J., **52** (1987) 807.
[17] J. Lauzeral, J. Halloy, and A. Goldbeter, Proc. Natl. Acad. Sci. USA, **94** (1997) 9153.
[18] Y.X. Li and A. Goldbeter, Biophys. J. **55** (1989) 125.
[19] G. Dupont and A. Goldbeter, BioEssays **14** (1992) 485.
[20] G. Dupont and A. Goldbeter, Biophys. J. **67** (1994) 2191.
[21] R.J. Konopka and S. Benzer, Proc. Natl. Acad. Sci. USA, **68** (1971) 2112.
[22] P.E. Hardin, J.C. Hall, and M. Rosbash, Nature, **343** (1990) 536.
[23] B.C. Goodwin, Adv. Enzyme Regul., **3** (1965) 425.
[24] A. Goldbeter, Proc. Roy. Soc. London Series B **261** (1995) 319.
[25] J.C. Leloup and A. Goldbeter, J. Biol. Rhythms, **13** (1998) 70.

SECTION III. DEVELOPMENT OF THE CENTRAL NERVOUS SYSTEM

COURSE 8

AN OVERVIEW OF NERVOUS SYSTEM DEVELOPMENT

Olivier Pourquié [a] and Franck Bourrat [b]

[a] *Developmental Biology Institute of Marseille (IBDM), LGPD-UMR CNRS 6545,
Campus de Luminy, Case 907, 13288 Marseille Cedex 09, France*
[b] *Laboratoire de Génétique des Poissons, INRA, Domaine de Vilvert,
78 352 Jouy en Josas Cedex, France*

G. Zaccai, J. Massoulié and F. David, eds.
Les Houches, Session LXV, 1996
De la Cellule au Cerveau
From Cell to Brain: Intra- and Inter-Cellular Communication –
The Central Nervous System

Contents

1. Neurogenesis and neural differentiation

1.1. Cell production and cell differentiation in the nervous system

Once the neural primordium has achieved a tubular shape, its cells undergo a phase of massive proliferation. All the glial and neuronal cells of the mature nervous system will ultimately be produced by this process (i.e., no cells will be recruited from outside).

The pattern of cell proliferation taking place in the neural tube is different from that (or those) taking place in other embryonic structures. The neural cells (or neurons) of the mature nervous system are also particular in that they lose, at some point of their developmental history, the ability to proliferate. Thus, after proliferation, a phase of differentiation produces the various neuronal cell types, depending both on the lineage of the cells (from which progenitor they come from) and on their interactions with their neighbours.

1.2. The proliferative zones: the ventricular neuroepithelia

At the end of the neurula stage, the central nervous system (CNS) assumes the shape of a tubular, pseudo-stratified neuroepithelium. The anterior, or cephalic end of this tube is slightly swollen (Fig. 1a). Its lumen will ultimately give rise to the various brain ventricles and, in the spinal cord, to the central canal. These ventricular cavities are filled with the cerebrospinal fluid, which circulates slowly in them, providing support to the brain and, to some extent, cushioning it from shocks.

1.2.1. The tritiated thymidine method
The cells lining the lumen of the neural tube are arrayed to constitute a pseudos-tratified neuroepithelium (Fig. 1b). They are mitotically extremely active.

The exact pattern of cell production in the vertebrate neural tube has been elucidated by the use of the tritiated thymidine method. This technique, the impact of which has been truly impressive in the field of developmental neurobiology, consists in providing the embryo, or, in mammals, the pregnant female, with radioactive thymidine (Thy*). This molecule is integrated in the DNA of the cells undergoing mitosis (and only these ones) at the time when Thy* is given. These

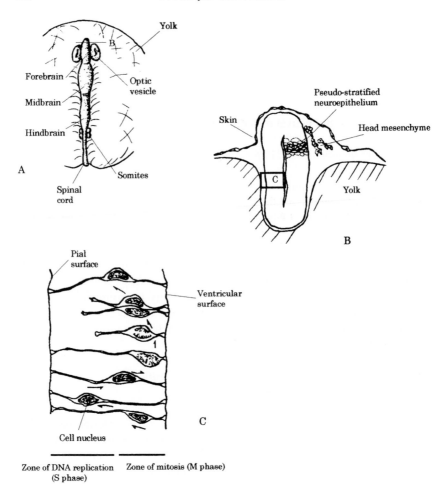

Fig. 1. Neuronal proliferation.(A) Schematic drawing of a fish (medaka, stage 19) embryo in dorsal view. (B) Transverse section at the level indicated in (A). The wall of the neural tube is made of a pseudo-stratified neuroepithelium. (C) Schematic drawing of the area corresponding to the box in (B). The nuclei of the cells move up and down. The DNA synthesis (S phase of the cycle) occurs while the nuclei move towards the pia, and the M phase occurs near the ventricular surface.

cells are thereafter visualised by autoradiography on histological sections. Due to the rapid turnover of the thymidine pool, the labelled molecule remains available for a short period of time (a few hours in mammals) to the embryos.

1.2.2. Pattern of neuron production

When used in early embryos, this method (along with other techniques) has shown that the mitosis takes place near the lumen of the neural tube, whereas the DNA synthesis (and therefore the incorporation of Thy*) happens in the outer portion of the wall, also called the mantle layer (Fig. 1c). It should be noticed that this results from a migration of the nucleus of the neuroepithelial cells, which extend as a long and thin cytoplasmic procession attached to the outer and inner surface of the neural tube, a pattern of intracellular migration that has been known for quite a long time (Sauer, 1935).

1.2.3. Neuronal birthdates

The Thy* method has wider uses, and is especially efficient to plot the final position of neurons generated at a given developmental period. Indeed, if the cell that has incorporated the Thy* continues to divide, each mitosis will dilute the tracer (semi-conservative replication of the DNA), until the labelling becomes no longer visible. Therefore, the Thy* technique is best suited to identify the cells undergoing their last round of mitosis. As already stated above, the neurons have lost their ability to divide, which, in other words, means that the time at which they underwent their last mitosis can be precisely defined. The term "birthdate" is commonly used to refer to this developmental event.

As will be exemplified later, in most if not all the nervous structures the neurons occupying a given position are generated during a given, and precise, time window.

For the moment, suffice it to say that which precedes is true for the neurons. The glial cells, which are generated later, are different in that most of them keep the ability to divide, and are therefore not definitely postmitotic. Although the glial progenitors are indeed produced in the ventricular neuroepithelium, glial cells can proliferate in all parts of the developing CNS, as well as in the adult, notably in response to an injury.

1.3. Cell lineage in the CNS

1.3.1. The general problem of cell lineage

The lineage of a single progenitor cell is defined as the cells it will eventually give rise to. A progenitor is said to be restricted, or committed, if it produces only a specific kind of cells (muscle, a given type of neurons, ...). It is said to be pluripotent if it gives rise to different kinds of cells. This is, obviously, a time-dependent problem, as cell lineages become more and more restricted as development proceeds: at one extreme, the single egg cell is perfectly totipotent, and at the other the last mitosis of a progenitor seldom produces widely different cells (but see below). The important problem of cell lineage is especially relevant in the CNS, which is made of literally thousands different cell types. Is the neural tube of early vertebrate embryos a mosaic of committed neuronal and glial progenitors, or do the cells of the ventricular neuroepithelium remain totipotent until late in development?

To answer this question, one must be able to identify the progenitors of a cell population, to individually label them, and to follow their successive mitosis up to their final, differentiated state. Finally, one has to challenge identified progenitors by transplanting them in a different cellular environment, and following their fate in this new context. Several methods have been used for this purpose.

1.3.2. Cell lineage in Caenorhabditis elegans

Given the complexity of the vertebrate CNS, the study of cell lineage in simpler systems has often been favoured by developmental neurobiologists. The better known animal in this respect is the nematode (round worm) *Caenorhabditis elegans*, the study of which has been pioneered by S. Brenner and collaborators (Brenner, 1974). We shall briefly present this work, to exemplify the kind of analysis that can be performed in this field.

The adult of this small (~1 mm) worm, which lives in the humus of forests where it feeds on bacteria, is made of a fixed number of cells: 959 somatic cells, out of which 222 are neurons (incidentally, this is amongst the highest "brain/body" ratio of all known organisms!). The transparency of *C. elegans* embryos and adults has allowed to trace the lineage of each individual cell at each developmental stage (Sulston et al., 1983), just by looking at the live embryo under a microscope equipped with interferential contrast ("Nomarski") optics. These analyses have unravelled that not only is the final number of cells invariant from one worm to the other, but so is the developmental sequence leading from the single egg cell to the adult. Moreover, in most cases the fate of a progenitor cell (i.e. the kind and type of cells it gives rise to) does not depend on its environment, and remains identical if it is transplanted elsewhere in the embryo, or if some of its neighbours are destroyed (by laser illumination).

For example, one of the two cells produced by the first division of the egg will ultimately give rise to 214 out of the 222 neurons of the adult worm, but also to other cell types (muscle cells for instance). Examples are also found where a cell's last division produces a neuron on one hand, and a muscle cell on the other, a situation never encountered in vertebrates where the mesodermic (muscle) and ectodermic (neuron) lineages diverge early and remain separate.

Although the simplicity of organisation of the CNS of this worm, and the availability of extremely powerful molecular and genetic tools, have permitted in depth analysis of the determinants of cell lineage and fate at the single cell level, it is clear that all the results cannot be extended to vertebrates, which, by all accounts have a much less rigid mode of development. They are also answerable to different kinds of studies.

Table 1

Neuron numbers in animals

Animal	Neuron number
Nematodes (small round worms)	300–400
Annelids (earthworms, leeches)	10^4 to 10^5
Insects	10^6
Octopus, reptiles, smallest mammals	3×10^6 to 5×10^6
Humans	85×10^9
telencephalon (forebrain) (cerebral cortex mainly)	12×10^9 to 15×10^9
cerebellar granule cells	70×10^9
brainstem, spinal cord	10^6
Whales, elephants	200×10^9

1.3.3. Cell lineage studied by injection of intracellular tracers

With some exceptions (embryos of many fish species are a case in point) most vertebrate embryos are opaque; moreover, the huge number of cells in their nervous system makes any systematic ("*C. elegans*-like") analysis impossible. Orders of magnitude of neuron numbers in the brains of various animal species are given in Table 1.

Therefore, other strategies have been developed for the study of cell lineage. In oviparous vertebrates, such as birds and amphibians (salamanders, newts, frogs and toads) the embryo is amenable to experimental manipulations. The method of choice consists in these cases in injecting a single cell with a high molecular weight tracer. This latter point is of importance, since embryonic cells quite often communicate with each other though intercellular junctions ("gap junctions") which allow diffusion of small molecules. Thus, use of "big" molecules is requested to ensure that the tracer does not leak into neigbouring cells. Another mandatory requirement is, of course, that the tracer should be "neutral", i.e. does not interfere with cellular metabolism, and still another one is that it should remain detectable even after the dilution of consecutive mitosis.

Two kinds of molecules have been widely used: dextran polymers linked to fluorophores (fluorescent tracers, that can be directly visualised) and enzymes, the most important being a peroxidase isolated from a plant, the horseradish (HRP). In this latter case, the tracer is visualised by an histochemical reaction carried out on tissue sections.

For example, with this latter tool M. Jacobson and collaborators have mapped the lineage of all the cells (or blastomeres) of the Xenopus blastula (for a review, see Jacobson, 1985). They were able to demonstrate that, from stage 512 cells onwards, the embryo is divided in compartments corresponding to the future CNS

territories. In other words, a blastomere injected at this stage will ultimately produce cells (neurons) belonging to one, and only one, CNS region (ventral spinal
cord, or telencephalon, ...). However, the situation is quite different from that encountered in *C. elegans*. Firstly, the relationship between a progenitor and the cell
population it gives rise to is probabilistic, i.e., the descendants of a blastomere
will differ from one embryo to the other, although they will always contribute
to the cell population of the same territory. And secondly, this spatial restriction
does not appear before the 512 cell stage. Tracer injections in earlier blastomeres
will not label specific sub-regions of the CNS. However, if HRP is injected in
one of the two blastomeres of a two cell-stage Xenopus embryo, the CNS of the
resulting larva will be labelled on one side (left or right) only, with the exception
of the retina and ventral diencephalon, where labelled and non-labelled cells are
intermingled (Jacobson and Hirose, 1978).

The same kind of analysis has been carried out in the zebrafish, with fluorescent
tracers, taking advantage of the remarkable transparency of this fish's embryo (see,
e.g., Kimmel and Law, 1985).

1.3.4. Cell lineage studied with engineered retroviruses

Retroviruses are RNA viruses which integrate themselves in the genome of infected cells. Genetically engineered retroviruses, derived from viruses infecting
birds and mammals, have been produced, in which the gene responsible for cell
toxicity (in most cases, an oncogene) has been replaced by a reporter gene, typically that coding for β-galactosidase, an enzyme easily detectable by a histochemical method. Moreover, these virus have been modified so that they retain the
ability to integrate themselves in the genome of the host cell, but lose the capacity
to replicate normally (and infect other cells). Therefore, these modified viruses
become stable components of the infected cell DNA, being replicated when and
only when the cell itself divides.

They are valuable markers of cell lineage, because, contrary to the above-mentioned injected tracers, the label is, by construction, not diluted at each mitosis.

Such engineered viruses are used at very low titers (highly diluted solutions)
to ensure that, statistically, only isolated cells will be infected. Indeed, if two
or more cells in close vicinity are infected at the same time, their descendants
will be impossible to sort out. Such viral solutions are then injected in embryos:
for CNS studies, the injection is typically made in the ventricular spaces, thus
leading to infection of neuroepithelial cells near the injection point. The virus is
transmitted to the descendants of the infected cells, and the reporter gene allows
to visualise the lineage derived from the infected cell. It should be noticed that
the viral infection is a random process, and therefore there can be no control over
what kind of progenitor is infected.

Despite this major drawback, these methods have allowed to perform lineage studies in the CNS of animals (primarily mammals) that are otherwise not amenable to such analysis (for a review of this technique, see Cepko, 1988).

Although these techniques have not yet been widely used, they have allowed to demonstrate that, generally speaking, the progenitor cells in the vertebrate CNS are pluripotent until late in development: they often remain able to produce different neuronal types at their final mitosis.

1.3.5. Other techniques for the study of cell lineage in the CNS
Two other techniques, already mentioned, can be used: one involves the transplantation, or grafting, of cells from one species to another. The transplanted cells, and their descendants, should remain identifiable. One can thus follow their lineage, if progenitors are transplanted, and especially if the transplantation involves single cells, which can be technically difficult.

Finally, cell lineage can be followed in vitro: this is achieved by isolating single progenitors, and growing them in a defined medium culture (clonal cultures). One can therefore test some of the factors that govern neural differentiation. This approach has been particularly useful to analyse so called stem cell populations, which are progenitor cells persisting, in some cases, in the adult brain.

1.4. Neuronal differentiation

1.4.1. Neuronal phenotype
The phenotype of a neuron, as of any cell in a multicellular organism, is defined by a set of criteria amongst which its shape, its localisation, and so on. More specific, and more relevant for these cells specialised in treating and transferring information are their connections (from which neurons it receives inputs, and to which it sends outputs), and the kind of transmitter it uses. For example, motoneurons are amongst the very few neurons using acetylcholine; or, in the peripheral nervous system (PNS), only sympathetic neurons use noradrenaline.

Immunocytochemical techniques are a simple and efficient tool for characterising the transmitter phenotype of a neuron. Antibodies are indeed available that recognise either the transmitters themselves, or key enzymes in their synthesis. For instance, antibodies against tyrosine hydroxilase, a critical enzyme in the catecholamines synthesis pathway, allow the detection and identification of neurons using these transmitters (dopamine, noradrenaline).

1.4.2. Factors determining neuronal phenotypes
The analysis of cell lineage in the CNS is obviously to be undertaken with the question of phenotype determination in mind. In other words, the central question here is to understand whether, and to what extent, the phenotype of a neuron is determined by its ancestors (its lineage), and/or by the interactions with its

neighbours. The answer is obviously a complex one, and depends largely on the animal model studied.

As mentioned above, the ultimate test is to transplant a progenitor in a different cellular environment, and to compare the lineage it gives rise to in this new situation to the "normal" one. It is important to keep in mind that, by "different environment", one can mean not only different cells, but also the same cells at a different developmental stage (so called "heterochronic" grafts, where, for instance, a piece of early embryonic tissue is transplanted in an older embryo, but at its "normal" place).

We saw previously that in *C. elegans*, the phenotype of a neuron is largely, if not exclusively, determined by its lineage, i.e., by the progenitor it comes from. The vertebrate situation is different, and indeed exhibits a greater degree of plasticity. We shall give a few examples:

Pioneer studies performed in the 70's by Teillet and Le Douarin on the differentiation of neural crest derivatives (Le Douarin et al., 1975). After replacing, in bird embryos, the neural crest fated to give rise to the enteric nervous system (ENS, the neurons of the viscera, which are cholinergic) with the neural crest from an area normally giving rise only to sympathetic and sensory neurons (which have an adrenergic phenotype), the ENS of the grafted animals is populated with cholinergic neurons coming from the graft.

This experiment can be interpreted in two ways: either the neural crest, at the time of grafting, is made of one population of pluripotent progenitor cells, and the choice between adrenergic or cholinergic lineage will be dictated by their environment; or, the neural crest is an heterogenous population of progenitors already committed to one lineage or another, and the environment will determine which progenitors will survive.

In vitro experiments carried out by Patterson and collaborators (see, e.g., Chun and Patterson, 1977) have demonstrated that the first interpretation is correct: the lineage choice is indeed made at the level of individual pluripotent progenitors, and is determined by environmental factors, several of which have been identified (steroid hormones, growth factors such as the bFGF or NGF, etc.).

Heterochronic transplantations experiments have been carried out, for example, by the group of McConnell (McConnell and Kaznowski, 1991; reviewed in McConnell, 1995) on the developing neocortex of the ferret. As will be seen in the next section, the cerebral cortex of mammals is built up by an "inside-out" migration of neurons, the oldest cells populating the deep cortical layers, and the youngest (late-generated) cells making the external layers. Progenitors have been labelled with tritiated thymidine, transplanted into older host brains, and the laminar fate of the daughter cells assayed at the end of the migratory period. It has been shown that this fate is dependent upon the phase of the cell cycle the progenitors were in at the time of transplantation. Descendants of cells transplanted when in

S-phase migrate as the neurons of the host (i.e. to the external layers, which is for them a change in fate). But if the progenitors are transplanted later in the cell cycle, they produce neurons which populate the deep layers (which correspond to their normal fate). Thus, environmental factors are important in determining neuronal fate, but intrinsic properties of the neuronal progenitor do play a part (in this case, a cyclical ability to respond or not to environmental cues).

Finally, one phenomenon should not be underestimated in these transplantation experiments: it has been called "community effect" by Gurdon (Gurdon, 1988; Gurdon et al., 1993), and results in the fact that, to really challenge a transplanted cell, one must ideally transplant it alone; only in this situation is the cell confronted with a new environment. If solid groups of cells are transplanted, these cells, or at least some of them, keep their previous neighbours, and therefore their "old" environment, complicating the interpretation of the experiments.

1.5. Conclusion

After neurulation, a phase of intense proliferation takes place in the neural tube. The neural progenitors divide according to a specific pattern in the wall of the tube, with is made of a neuroepithelium. This structure first produces neurons, the birthdate of which can be determined by the tritiated thymidine method. Cell lineage studies allow to trace the developmental history of neurons, and to unravel some factors determining their phenotype.

In the nematode *C. elegans*, cell fate appears to be for the most part lineage-dependent, and thus intrinsically programmed. In vertebrates, techniques such as injections of intracellular tracers, or of genetically engineered retroviruses, allow to follow cell lineages in much more complex systems. Contrary to the nematode situation, cell fate in the vertebrate CNS is often not determined until late in development, and depends heavily on environmental factors.

2. Neuronal migration and axon growth

After their final mitosis in the ventricular neuroepithelium, postmitotic neurons of the vertebrate CNS usually migrate towards their final location. During, or immediately after, this migratory step, they extend cytoplasmic processes, called neurites, which actively seek their targets (other neurons or muscles). Therefore, as a result of these cellular movements, the scaffold of intricate neural connections which characterises the CNS becomes apparent.

2.1. Neuronal migration

Contrary to the neural crest cells, which invade many different tissues in the embryo, neuronal migration is restricted to the CNS.

2.1.1. The two main types of neuronal organisation in the vertebrate CNS

Topographically, the adult vertebrate CNS (brain proper, and spinal cord) is an extremely ordered structure (for a general presentation of the vertebrate CNS anatomy, see Butler and Hodos, 1996). Two main kinds of organisation are encountered: in the first one, the neurons are arrayed in layers. In these structures, called cortices, the layers populated with cell bodies often alternate with fibrous layers, made of neuronal processes (axons and/or dendrites), and devoid of neuronal somata, although glial cell bodies are present. This cortical, or laminar, type of organisation is more frequently found in the dorsal parts of the vertebrate brain. Examples include the cerebral cortex of mammals, the optic tectum of lower vertebrates, and the cerebellar cortex in all vertebrates.

In the second type, the neurons are grouped in cellular masses called nuclei. In neuronal nuclei, which are usually bilaterally arrayed with respect to the midline, the neurons may or may not be of the same type. Examples of such structures (called nuclear structures) include motor and sensory nuclei of the hindbrain and midbrain, and the thalamus.

The boundaries of neuronal nuclei can be sharp and well delineated, but also rather loose, as is the case for the reticular formation of the hindbrain.

In all types of structures can be found so called "projection neurons", which connect with other structures, close or remote, and "interneurons", the axons of which connect locally with other neurons in the same nucleus. It has been postulated that the increase in brain size and complexity in vertebrates is largely due to an increase in interneuronal populations (i.e., an increase in the "local" treatment of information).

We have seen in the previous section that all neurons, whatever their kind, are generated in the vicinity of the ventricular space, the lumen of the neural tube. But very rarely does the adult CNS in vertebrates keep the shape of an hollow cylinder (some primitive fishes are a partial exception). Therefore, the adult neuronal topography implicates that cellular movements take place during CNS morphogenesis, and that neurons migrate from the neuroepithelium towards their final location.

2.1.2. Radial migration and morphogenesis of cortical structures

This is probably the most important kind of cell movement in the vertebrate CNS. It has been beautifully described by the great spanish neuro-anatomist Ramon y Cajal at the turn of the century (Ramon y Cajal, 1911). In the 60's and 70's,

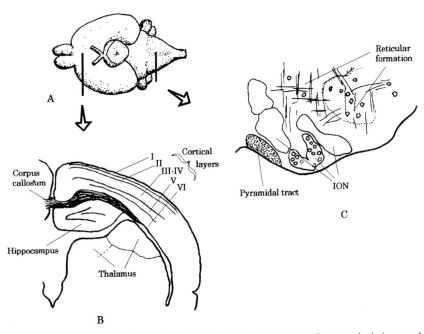

Fig. 2. Neuronal organizations in vertebrate CNS. (A) Schematic drawing of a mouse brain in ventral view. (B) Cortical (laminar) organization of the cerebral cortex. The neurons and fibers are disposed in six layers, numbered from I (most superficial) to VI (deepest). (C) Nuclear organization in the hindbrain. The neurons and fibers are disposed in well-defined nuclei (like the inferior olive, ION) and tracts (like the corticospinal, or pyramidal, tract), or are more loosely organized, like in the reticular formation.

finer details have been added, by the use of tritiated thymidine first (see, e.g., Angevine and Sidman, 1961), and by the works of P. Rakic (Rakic, 1971, 1972), who studied radial migrations in monkey cerebral cortex and cerebellum with electron microscopic techniques.

In the cerebral cortex of mammals, newly generated ("born") neurons move on a substratum made of specialised cells, called radial glia (Fig. 2). These cells are arrayed radially, perpendicular to the long (antero-posterior) axis of the neural tube, and exhibit a very elongated shape, with both ends attached respectively to the ventricular and external (pial) surfaces. Thus, their cell bodies extend all across the wall of the developing neural tube. It is a transient cell population, which exists only during the developmental stages, and most of radial glia cells later differentiate into astrocytes.

These radial glia cells act as guides for the migration of neurons generated in the ventricular neuroepithelium. The neurons stop their movement at the exter-

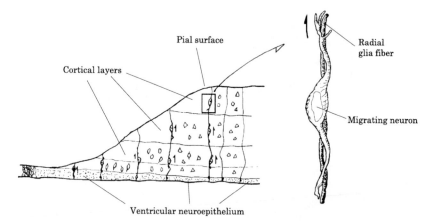

Fig. 3. Formation of cortical layers by radial migration. Neurons generated in the ventricular neuroepithelium migrate "upwards", guided by radial glia fibers. This pattern of radial migration ("inside-out") is found, for example, in the mammalian cerebral cortex.

nal end of these cells, which means that they accumulate successively further and further away from the ventricle. Thus, the first generated, or oldest cells, are located near the ventricle and make the deep layers of the cortex, whereas the latest generated cells form the superficial layers.

This pattern of neuronal migration, often referred to as "inside-out", operates in the ontogenesis of several cortical structures (cerebral cortex, hippocampus, etc. in mammals. In the cerebellar cortex, some cell layers (notably the Purkinje cell layer) are also set up by this mechanism. In addition, a secondary neuroepithelium forms later in development, and covers progressively, in a posterior to anterior sequence, the external surface of the cerebellar anlagen. This neuroepithelium, called the external granular cell layer, is active for a long time after birth in most mammals and generate neurons – the granule cells, by far the most numerous neurons of all vertebrate CNS – which migrate also radially, but in an opposite direction ("outside-in", see Fig. 3). They use as substratum and guide a special type of radial glia, called Bergman glia, and end their migration near the ventricular surface to make the granular cell layer of the cerebellar cortex.

Study of mutant mice (*weaver* mutants) has shed some light on the interactions between radial glia and migrating granule cells. In these mutants, radial glia is abnormal, granule neurons do not migrate and ultimately degenerate. It is difficult to ascertain if this is due to a defect in the granule neurons, or in the glial cells. It is possible to purify populations of radial glia cells on one hand and of granule neurons on the other hand. The migration of granule neurons on glial substratum

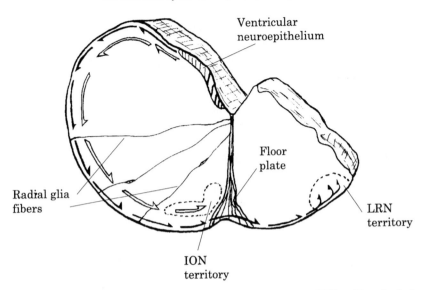

Fig. 4. Migration of precerebellar neurons. The neurons of the inferior olive (ION) and lateral reticular nucleus (LRN) are generated in the dorsal ventricular neuroepithelium of the hindbrain. They migrate ventrally (LRN: filled arrows; ION: open arrows) to their final territories. This migration is not guided by radial glia fibers. The floor plate may attract these migrating neurons.

can then be reproduced in Petri dishes. In vitro experiments have shown that wild-type neurons can migrate on weaver glia, but not vice versa (Hatten et al., 1986). This clearly indicates that the primary effect of the mutation is neuronal, as has been later confirmed.

2.1.3. Other patterns of neuronal migration

Some nuclear structures in the vertebrate CNS are set up by mechanisms different from radial migration. A case in point is the ontogenesis of the inferior olive and lateral reticular nucleus, studied by Bourrat and Sotelo (1990). These are bilateral nuclear structures, located in the ventral part of the posterior hindbrain. Both are pre-cerebellar structures, meaning that the targets of their axons are cerebellar neurons (the Purkinje cells for the inferior olive and the granule cells for the lateral reticular nucleus).

The neurons that make these two nuclei are generated in the ventricular neuroepithelium of the hindbrain (see Fig. 4). They thereafter migrate circonferencially in a ventral direction, some ultimately crossing the midline (the lateral reticular neurons), the other not (the olivary neurons). The mechanism of this migration must be different from the above-mentioned radial migration, because, although radial glia does exists in the developing hindbrain, the neuronal movements are in this case perpendicular to it.

Another, quite spectacular example of neuronal migration is provided by the GnRH (gonadotropin releasing hormone) neurons. These neurons are, in the adult mammal, located in the hypothalamus. However, they are generated in the olfactory placode, the most anterior part of the central nervous system, and the structure which gives rise to the odour-sensitive neurons of the nasal neuroepithelium. During ontogenesis, the GnRH neurons migrate along the developing olfactory nerve, penetrate in the CNS and ultimately reach the hypothalamus (Schwanzel-Fukuda and Pfaff, 1989).

In the Kallmann–de Morsier syndrome, these neurons, as well as the neurons of the olfactory neuroepithelium, are missing. The consequences are a lack of sensibility to odours, and a abnormal development of the gonads (due to the absence of GnRH neurons).

2.2. Axonal migration

2.2.1. The growth cone, motor of axonal elongation

The most unusual shape of a neuron, with its extremely long dendrites and axon, is built in the course of ontogenesis by the outgrowth of cytoplasmic processes. The axonal process, in particular, grows until it reaches its target (which might be another neuron or a muscle cell) with which it establishes a synaptic contact.

Ramon y Cajal was the first to describe the specialised structure located at the tip of the growing axons and dendrites and to recognise its pivotal role in neuritic elongation. As soon as the neurites began to protrude from the neuronal cell body, they exhibit a *growth cone* at their tip. The growth is achieved by addition of membrane at this level. Thus, axons and dendrites do not grow, as is the case for hairs, by adding new materials at their proximal (basal) end, neither all along their length, but by specific addition of material at their tip. The materials are indeed synthesised in the cell body of the neuron, then carried to the growth cone by the fast and/or the slow axonal transport. The direction of growth is dictated by interactions between the growth cone and its environment (direct interactions with the substratum, presence of soluble factors, electric fields, etc., see below). The growth cone is the only mobile structure of the developing neurite.

Morphologically, growth cones are quite diverse (see Fig. 5). They usually possess two types of mobile protrusions: filopodia, long and very thin cytoplasmic tubes, and lamellipodia, much larger structures resembling fans. The shape of growth cones can change according to its environment. It is usually assumed that, when neuritic growth is fast and simple (along a pre-existing nerve for example) the growth cone has a very simple shape (more or less that of a paintbrush Fig. 5a). In contrast, when the growth cone is faced with many directional choic-

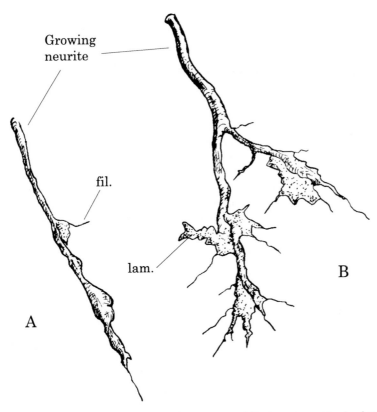

Fig. 5. Morphology of growth cones. The growth cones are specialized structures at the tip of growing neurites. They can have "simple" (A) or more complex shapes (B) depending on their environment. Numerous filopodia (fil.) and lamellipodia (lam.) protrude from the cone.

es, it assumes a complex morphology, with several lamellipodia and numerous filopodia (Fig. 5).

The neurite behind the growth cone is stabilised by the layout of a cytoskeleton, mostly made of microtubules. These structures are polymers of α- and β-tubulin dimers in the same orientation. In axons, elongation of this microtubule cytoskeleton takes place only in the growth cone.

The active behaviour (changes of shape, protrusion–retraction of filopodia, etc.) of the growth cone itself is due to a cytoskeleton of numerous actin microfilaments. Actin is present as either globular actin (actin G) or fibrillar actin (actin F). The network of actin filaments is linked to the microtubules and to the membrane by specialised proteins.

2.2.2. Mechanisms of neuritic elongation

Following the pioneer in vitro studies of Harrison (1910) at the turn of the century, analysis of factors at play in directional choices made by growth cones has unravelled that several mechanisms collaborate in that process.

Stereotropism. The first one is called stereotropism, and characterises the "natural tendency" of growth cones to navigate along well-defined physical borders, such as scratches at the bottom of a Petri dish. In the embryo, this mechanism could be used by growth cones migrating in particular extracellular spaces (which are considered to be wider in embryonic than in adult tissues). Indeed, this hypothesis has been formulated in an elaborated form ("blueprint hypothesis"), which postulates that kinds of "tunnels" do exist in the wall of the neuroepithelium, or in other embryonic tissues, which delineate the pathway of some growth cones. However, this hypothesis alone appears inadequate to explain the exquisite specificity of neuronal connections.

Differential adhesiveness. A second mechanism relies on differential adhesiveness of the growth cones substrata. The extracellular matrix, which is indeed abundant in embryonic tissues, is made of several kinds of molecules, some of which are excellent substrata for growth cone elongation, as assayed by in vitro experiments. Laminin, for example, belongs to this category. It is a very big protein (850 kDa), composed of three chains A, B1 and B2. Laminin is a component of basement membranes, and is not present in the adult CNS. However, during development, laminin can be transiently found along several axonal pathways (the optic nerve, for instance), and therefore could contribute to the preferential elongation of growth cones along these routes. Laminin is recognised through the binding with a receptor of the integrin superfamily, which is also transiently expressed at the surface of some growth cones. Integrins are dimeric molecules composed of a β chain, more or less conserved between the molecules of this family, and an α chain, highly variable, and responsible for the binding specificity of the molecule. These molecules comprise an extracellular domain implicated in calcium-dependent adhesion phenomena, and an intracellular (cytoplasmic) domain linked to the actin cytoskeleton (see above). Most integrins recognise molecules of the extracellular matrix such as laminin, fibronectin, or collagen.

Laminin is certainly not the only molecule at play: indeed, many other molecules of the extracellular matrix, such as fibronectin, or several glycosaminoglycans, are very good in vitro substrata for growth cone extension. Moreover, the differential adhesiveness hypothesis is not restricted to extracellular matrix. Cellular and neuritic surfaces can also make favourable substrata, notably by expressing adhesion molecules such as the N-CAM or some cadherins. When these molecules are used in vitro as substrata, they are able to greatly stimulate neuritic extension. This indicates that, in addition to their role in regulating adhesion between the growth cone membrane and its substratum, they might be involved in transducing signals from the environment to the neuron.

Another important aspect of these differential adhesiveness mechanisms lies in the fact that the interactions between growth cones and substrata are not always "positive". Indeed, several "repulsive" molecules, such as the tenascin (an extracellular matrix molecule) have been characterized: they block neuritic extension, instead of favouring it. Another example is the collapsin, which induces a retraction of growth cones which contact it. The molecules of this class are usually expressed transiently during CNS development, and might explain the presence of impassable barriers, or boundaries, that growing axons never cross (for example, the growing axons coming from the retina – which form the optic nerve – never invade the telencephalic territory, but always home towards the dorsal mesencephale).

Electric fields (galvanotaxis).   A third guidance mechanism relies on the presence of weak electric fields in embryos. This hypothesis has been formulated long ago by Kappers, but only recently have these fields be shown to exist, in vivo and in vitro (Jaffe and Stern, 1979). In this latter situation at least, growth cones of cultured neurons are able to orientate and grow along these fields (they migrate towards the cathode, see Jaffe and Poo, 1979).

Chemotaxis.   Finally, a fourth mechanism, hypothesised one century ago by Ramon y Cajal, is chemotaxis. Several cell types (bacteria, some blood cells, some amoebas, etc.) are able to recognise a gradient of soluble molecule(s), and to migrate towards or away from its source. At the end of the 70's, the same kind of mechanism has been shown to be at play in growth cones. In vitro, a source of nerve growth factor (NGF) induces from a distance a re-orientation of neuritic elongation towards it, for some neuronal types. However, such phenomena appear to play little if any role in physiological (in vivo) conditions, and it is doubtful that neurotrophins (the family of molecules to which the NGF belongs) are implicated at all in growth cone guidance.

Very recently, the group of Tessier-Lavigne (Kennedy et al., 1994; Tessier-Lavigne and Placzek, 1991) has characterised a family of "truly" chemotactic molecules, and has demonstrated their direct involvement in axonal guidance. These molecules are the netrins, one of which is produced by the floor plate in the embryonic neural tube. This protein attract the growth cones of a population of spinal cord neurons (the commissural neurons), which migrate towards the floor plate.

2.3. Conclusion

In the vertebrate CNS, a majority of neurons migrate radially, from the ventricule to the external surface, using specialised cells, the radial glia, as substratum and guide. This migratory pattern accounts for the formation of laminar, or cortical, structures. Although it is the most important, radial migration is clearly not the only mechanism: in some cases, exemplified by the hypothalamic GnRH neurons, the neuronal migration takes place along axons. In other situations, such as the circumferential migration of the olivary neurons of the hindbrain, the precise mechanism at play is still unknown.

During their migration or shortly after, cytoplasmic processes (neurites) protrude from the neuronal cell body and growth towards their targets. A specialised structure located at the tip of the developing neurites, the growth cone, is responsible for their elongation and for the directional choices. These latter relies on several mechanisms, such as stereotropism, differential adhesiveness, response to weak electric fields, or chemotaxis.

References

[1] J.B. Angevine and R.L. Sidman, Nature **192** (1961) 766.
[2] F. Bourrat and C. Sotelo, J. Comp. Neurol. **294** (1990) 1.
[3] S. Brenner, Genetics **77** (1974) 71.
[4] A.B. Butler and W. Hodos, *Comparative Vertebrate Neuroanatomy. Evolution and Adaptation* (Wiley-Liss, New York, 1996).
[5] C. Cepko, Neuron **1** (1988) 345.
[6] L.L.Y. Chun and P.H. Patterson, J. Cell Biol. **75** (1977) 694.
[7] J.B. Gurdon, Nature **336** (1988) 772.
[8] J.B. Gurdon, P. Lemaire, and K. Kato, Cell **75** (1993) 831.
[9] R.G. Harrison, J. Exp. Zool. **9** (1910) 787.
[10] M.E. Hatten, R.K.H. Liem, and C.A. Mason, J. Neurosci. **6** (1986) 2676.
[11] M. Jacobson, Ann. Rev. Neurosci. **8** (1985) 71.
[12] M. Jacobson and G. Hirose, Science **202** (1978) 637.
[13] L.F. Jaffe and M.-M. Poo, J. Exp. Zool. **209** (1979) 115.
[14] L.F. Jaffe and C.D. Stern, Science **206** (1979) 569.
[15] T.E. Kennedy, T. Serafini, J.R. de la Torre, and M. Tessier-Lavigne, Cell **78** (1994) 425.
[16] C.B. Kimmel and R.D. Law, Dev. Biol. **108** (1985) 94.
[17] N.M. Le Douarin, D. Renaud, M.-A. Teillet, and G.H. Le Douarin, Proc. Natl. Acad. Sci. USA **72** (1975) 728.
[18] S.K. McConnell, Neuron **15** (1995) 761.
[19] S.K. McConnell and C.E. Kaznowski, Science **254** (1991) 282.
[20] P. Rakic, J. Comp. Neurol. **141** (1971) 283.
[21] P. Rakic, J. Comp. Neurol. **145** (1972) 61.
[22] Ramon y Cajal, *Histologie du Sytème Nerveux de l'Homme et des Vertébrés* (Maloine, Paris, 1911).

[23] F.C. Sauer, J. Comp. Neurol. **62** (1935) 377.
[24] M. Schwanzel-Fukuda and D.W. Pfaff, Nature **338** (1989) 161.
[25] J.E. Sulston, J. Schierenberg, J. White, and N. Thomson, Dev. Biol. **100** (1983) 64.
[26] M. Tessier-Lavigne and M. Placzek, Trends Neurosci. **14** (7) (1991) 303.

COURSE 9

G-PROTEIN COUPLED RECEPTORS: THEMES AND VARIATIONS ON MEMBRANE TRANSMISSION OF EXTRACELLULAR SIGNALS

Philippe Vernier

Institut Alfred Fessard, UPR 2212 C.N.R.S., F-91198, Gif-sur-Yvette Cedex, France

G. Zaccai, J. Massoulié and F. David, eds.
Les Houches, Session LXV, 1996
De la Cellule au Cerveau
From Cell to Brain: Intra- and Inter-Cellular Communication –
The Central Nervous System

159

Contents

1. Receptors are essential cell components

The plasma membrane of cells is made of a lipid bilayer which separates two very different worlds. The intracellular milieu is essentially constant and organized by energetic metabolisms whereas its surrounding environment is highly changing and variable. Although the situation is somewhat different for a bacteria living in a wild world and for a eukaryotic cells quietly embedded in a large metazoan organism, both unicellular (protozoan) and multicellular (metazoan) organisms have a general requirement of being able to sense pertinent messages in their environment. Each cell needs to modify its metabolism, but also to change its shape and to move in response to extracellular cues reaching its close vicinity. Most of the extracellular messengers which bear significant information for cells are hydrophilic compounds, the size of which is much too large to diffuse across the plasma membrane. Thus, membrane receptors are the molecules designed and selected during evolution to recognize extracellular messages and to transmit the relevant information inside cells. The process by which extracellular signals are transformed into intracellular messages of different nature is called "transduction" (Fig. 1).

Receptors are generally just one among multiple components of a molecule assembly devoted to cellular signalling. The cascade of interactions triggered by the binding of extracellular messengers to receptor culminates in a so-called cell "response". The cell fate will be oriented toward differentiation or different functional states, toward mitosis (cell division) or even toward programmed cell death (apoptosis) according to demanding circumstances, the significance of which is given by how receptors are activated. Therefore, receptors are one of the very few essential components of cells, that is, the means cells use to sense their surrounding world and to adapt to it. What is generally named cell regulation is nothing more than the adaptation of cell behaviour to a given situation. Receptors are at the heart of the cell regulation mechanisms. According to their central role in cell biology, receptors have been the matter of an enormous amount of research work from the very beginning of this century (Ehrlich, 1913; Kenakin, 1983). Physiologists and cell biologists have studied receptors from several different points of view, but no one more than pharmacologists have striven to analyze the role of receptors.

P. Vernier

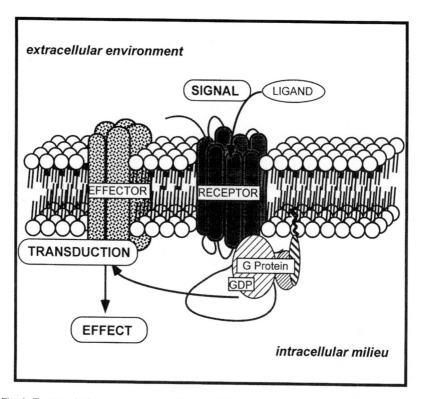

Fig. 1. To transmit the *signal* represented by the binding of an agonist ligand, the receptor needs
to interact with at least two other components: the heterotrimeric G protein and an effector protein
(enzyme, ion channel, ...). The conditional activation of the molecules constituting this *transmission
module* represents a *transduction* process which culminates in the so-called cellular *effect*.

Receptors have initially been discovered by pharmacologists as the entity required to bind drugs and toxins and to transmit their effects in organs. To be active,
drugs have indeed to bind in a saturable manner to receptors and then to interfere
with the effect of natural transmitters in cells and organs. For decades, receptors
were used mainly as operational concepts, the formalization of which allowed to
quantify drug activities. More recently, the isolation of receptors as membrane
proteins, the availability of their sequences owing to the power of genetic engineering and molecular biology, have prompted pharmacologists and biochemists
to analyze in detail the relationships between the structures and the activity of
receptors. This approach led to the surprising discovery of several hundredths
of different receptors existing in a single metazoan organism. A recent estimate
proposed that at least ten percent of the human genes are encoding membrane

receptors. However, the problem of studying such a high number of molecules appeared soon as presumably simplified by the fact all these receptors exhibited only a few prototypal structures. In many respects, the general principles of the organization of receptors and of their coupling with the set of molecules which initiates intracellular signalling pathways are highly conserved.

Receptors constitute superfamilies of membrane proteins, each characterized by a recognizable overall structure and a common way of activating intracellular signalling pathways. The main superfamilies of membrane receptors are those of ligand-gated ionic channels, membrane tyrosine kinases, janus kinase-coupled receptors, guanylate cyclase receptors and, last but not least, G-protein coupled receptors. This latter superfamily is by far the largest known family of membrane proteins. Even if the transduction mechanisms used by the various receptor superfamilies are significantly different, the principles of receptor function can be generalized from G-protein coupled receptors to the other receptor superfamilies. G-protein coupled receptors using biologically active monoamines are among the best known classes of such receptor molecules. In this essay, they will be used as examples of our current knowledge about the cell biology and regulation of membrane receptors. The structure–activity relationships and biochemistry of G-protein coupled receptor will be only briefly addressed*, the emphasis being placed on the broad questions of the receptor functions and regulation in the context of cell biology, on their biosynthesis and intracellular fate, on the complexity of their biological nature, and on their evolutionary origin. Finally, a limited number of references have been provided, obviously far from being exhaustive, but which may serve as a useful starting point for further research.

2. The mechanisms of signal transduction by G-protein coupled receptors

Hundredths of extracellular signals use G-protein coupled receptors as transduction devices to transmit corresponding information inside the target cells. It is enough to mention that many sensory informations such as light, odorants, pheromones, and more especially most of the small hydrophilic chemicals such as monoamines, nucleotides, ions, many peptidic hormones, neuromodulators or cytokines use receptors belonging to this family (Fig. 2). Moreover, these extracellular messengers have very often not only one but several receptor molecules. This is the case for biologically active amines which all share at least three different receptor subtypes in vertebrates. This class of compounds consists of well-known examples of extracellular messengers used generally as neurotransmitters.

---

* For this purpose the interested reader is invited to read recent reviews such as Gudermann et al. (1996), Valdenaire and Vernier (1997), Rohrer and Kobilka (1998).

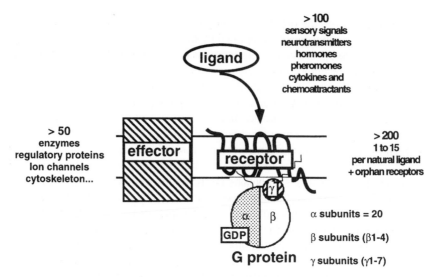

Fig. 2. The transmission module constituted by G-protein coupled receptors exhibits an extreme diversity of participating molecules. About a hundred natural ligands bind hundreds of receptors coupled to a few dozen G proteins which regulates dozens of effector proteins of different natures. Thus, the spatial and temporal organization of these molecular modules are crucial to confer to cells and organisms the required degree of specificity in signal transmission.

They are synthesized into neurones by a pathway of enzymatic modification of amino acids, such as tyrosine for the catecholamines dopamine, noradrenaline and adrenaline, or tryptophane for the indolamine serotonine (Vernier et al., 1993). Then, they are released outside the cells by exocytosis to act as modulatory neurotransmitters.

The transformation of extracellular signals which bind to receptors into intracellular information is achieved by the use of at least three interacting components: the membrane receptor itself, the G protein and an effector protein. The membrane receptor is a monomeric protein structurally characterized by the presence of seven transmembrane segments. The G protein is an heterogeneous multimer made of three subunits named α, β and γ. Since the α subunit is able to bind guanylic nucleotides (GTP and GDP) as well as to hydrolyze GTP, it give its name (G protein) to the heterotrimer. The β and γ subunits are tightly connected together so as to form a single functional entity. In animals such as mammals, more than twenty α subunits belonging to four classes have been identified (reviewed in Simon et al., 1991), that can be associated to five or six β and γ subunits, leading theoretically to a very large number of possible combinations.

The nature of the third component of this transduction module, the effector protein, is highly variable, as it can be as different as an enzyme (adenylyl cyclase, phospholipase C, phosphodiesterases, . . .), an ion channel (Na^+, K^+, Ca^{2+} voltage-gated channels, . . .), components of the cytoskeleton and several other regulatory proteins, the list of which is increasing every month. Given the tremendous multiplicity and diversity of the molecular components which settle the transduction mechanisms initiated by these receptors, it is clear that they fulfill a general necessity in cells and that they have encountered a large evolutionary success.

2.1. The activated receptors catalyze the G protein cycle

Our current view of the G-protein coupled receptors has been strongly influenced by the selectionist allosteric theory in which receptor molecules can spontaneously display different conformations. These conformations corresponds to different functional states of the receptors which each exhibit specific affinities for interacting components whether they are the so-called ligands or the coupled regulatory molecules. To be simple, it can be stated that the receptor oscillates between two functional states, a resting state and an activated state (Fig. 3). At the resting state, the G protein is associated to the cytoplasmic regions of the receptor in its heterotrimeric form, with a GDP molecule bound to the α subunit. In this situation, the effector protein is not functionally interacting with the receptor–G protein complex. According to the allosteric theory, a receptor in this resting state exhibits a high affinity for its natural agonist and it is stabilized by antagonist ligands.

In contrast, agonist ligands stabilize the active state of the receptor. This isomerization catalyzes a conformation change of the G protein α subunit, by decreasing the magnesium concentration required to promote the GDP–GTP exchange on the nucleotide site of the G protein. Indeed, since the GTP concentration in the cell cytoplasm is higher than that of GDP, the "opening" of the α subunit alone permits the exchange of GDP for GTP on the nucleotide binding cleft (Birnbaumer et al., 1990). The binding of GTP on the α subunit has dramatic consequences for the whole system, and represents the major transduction event in transmembrane signalling. The GTP-bound form of the α subunit dissociates from its cognate $\beta\gamma$ partners and loses contact with the membrane receptor. In contrast, the $\beta\gamma$ complex remains in the vicinity of the cytoplasmic loops of the receptor. The G protein dissociation triggers an allosteric change of the receptor binding site which decreases its affinity for the agonists, favouring them to leave the receptor binding site. In turn, the dissociated α and $\beta\gamma$ subunits each become able to activate a different target effector. Thus, a single round of receptor activation may modulate simultaneously two kinds of different effectors. The final cell response to the extracellular signal will be the consequence of these two separate pathways of intracellular signalling.

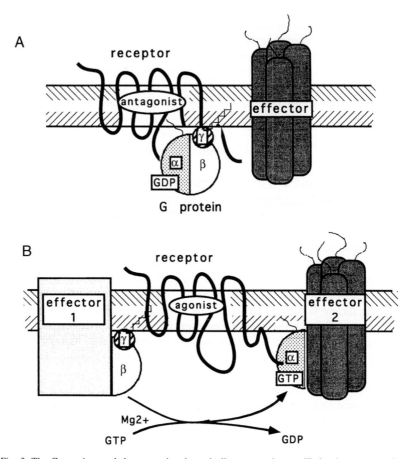

Fig. 3. The G-protein coupled receptor is schematically presented as oscillating between two functional states. In the resting state (A), the receptor is associated with the G protein in its trimeric, GDP bound form and it is stabilized by antagonist compounds. In the activated state (B), the receptor catalyzes the magnesium-dependent exchange of GDP for GTP on the α subunit of the G protein. This state is stabilized by agonist compounds and triggers the dissociation of the heterotrimer from the receptor. Both α and $\beta\gamma$ subunits are able to modulate the activity of different effectors.

The binding of GTP to the G protein α subunit is almost irreversible and this step gives the direction of the signal transmission (from the membrane to the cytoplasm), which is the most meaningful parameter of signal transduction. The only way for the G protein to stop the process is to hydrolyze GTP, owing to its intrinsic GTPase activity. As soon as GTP is hydrolyzed in GDP, the α subunit reassociates with the $\beta\gamma$ complex and the receptor, and the module is ready for a new cycle.

The rate-limiting step in receptor–G protein activation is GTP hydrolysis by the α subunit which is spontaneously rather slow (Bourne et al., 1990, 1991). Several cycles of G protein activation are generally performed in a time unit of receptor occupancy. The power and duration of the response will then depend on the activation time of the various subunits of the G protein. This observation accounts for the amplification of signal transmission which is another important characteristic of the receptor function. In this transduction process, overall, the receptor may be considered as the recognition component, the effector as the active component, whereas the G protein acts as a controller of the timing and of the direction of the reaction.

It is important here to stress the extraordinary role of GTPase in numerous aspects of the cell biology. In fact, as soon as some kind of control needs to be secured between an "upstream" component and a "downstream" component, cells generally use GTPase because of their property of conditional activation and de-activation (Bourne et al., 1990). The binding of GTP or GDP to the G protein is extremely stable and the transition from the GDP-bound state to the GTP-bound state, as well as the reverse action require the mediation of supplementary molecular components. The exchange of GDP for GTP can be either promoted by exchange factors or inhibited by guanine nucleotide release inhibitors (GNRI). In this respect, membrane receptors act as exchange factors for the heterotrimeric G proteins (Birnbaumer et al., 1990). In turn, GTP hydrolysis performed by G proteins is generally slow and may be accelerated by GTPase activating proteins (GAP). Thus the control of biological processes by G protein needs to fulfill two conditions. The first one is the GDP–GTP exchange and the second one is the GTP hydrolysis. By acting either on the nucleotide exchange or on the GTPase activity, regulatory components such as receptors are able to modify the duration and the efficiency of the transduction process.

2.2. *Structure–activity relationships in receptor–G protein coupling*

The membership of receptors, G proteins, and effectors to large protein families and the corresponding conservation of their interactions has been experimentally very useful. Indeed, as soon as a given mechanism has been elucidated for a given receptor, it can be often extrapolated to another kind of receptor, although a word of caution should be added before any generalization. In addition, the recent progresses of structural biology gave a boost to a huge amount of works which used site-directed mutagenesis and molecular modelling as means to better understand the receptor–G protein function.

As far as the receptor protein is concerned, its membrane topology has been deduced from previous work on bacteriorhodopsin (Henderson and Unwin, 1975; Grigorieff et al., 1996), even if this latter protein is not a G-protein coupled receptor (it is a proton pump). As for bacteriorhodopsin, the general topology of G-protein coupled receptors is particularized by the presence of seven hydrophobic stretches able to form α helices which span the plasma membrane (von Heijne and Manoil, 1990; Chothia and Finkelstein, 1990; High and Dobberstein, 1992). More recently, a closer look at the overall organization of receptors has been permitted on the basis of the high resolution structure of the visual rhodopsin, a bona fide G-protein coupled protein (Henderson et al., 1990). Based on these high resolution structures, it as been proposed that the seven amphipathic transmembrane helices are packed together in an anticlockwise manner, and that they are kinked in many instances, rendering very difficult any molecular modelization (Fig. 4b; Hoflack et al., 1994). The α helices are composed of polar residues which face each other on both sides of a dihedral cleft, whilst their hydrophobic surfaces contact directly the membrane lipids. The N-terminus of the receptor is extracellular and the C-terminus, generally anchored in the plasma membrane by myristylation, is intracellular. The first and second extracellular loop are most often linked by a disulphide bridge between two conserved cysteines (Fig. 4a).

A large number of mutagenesis studies have shown that a kind of pocket deeply embedded in the plasma membrane contains the binding site for most of the small ligand molecules which act as extracellular neurotransmitter. Extensive mutagenesis studies associated to powerful methods of molecular modelization have assigned a few residues located at the inner face of the helices to binding and transduction function (see for example Javitch et al., 1995; Hill-Eubanks et al., 1996). For example, amino acids located in the second and third helices have consistently been shown to participate in the formation of ion pairs with cationic groups present on small ligands, whereas residues in the fifth and sixth transmembrane segments are likely to coordinate hydroxyl groups of both agonist and antagonists ligands. In addition, some agreement begins to be reached on the delineation of the binding requirements for agonist and antagonist ligands, although significant differences may exist for some peptide receptors (Schwartz et al., 1995). In the case of the retinal bound to opsins or of monoamine ligands, the active state of the receptor is stabilized by compounds which interact with all the crucial residues of helices three, five and six. Conversely, antagonists are generally thought to prevent the access of agonists to the amino acid moieties responsible for receptor activation via their binding to a few residues of the fifth and sixth transmembrane helices (Fig. 4a).

Despite in these significant progresses in our understanding of the molecular mechanisms of drug activity, the precise mechanism which stabilize the active state of G-protein coupled receptor remain essentially elusive. The main clues about the way receptors are activated have been obtain by the surprising observation that substitution of residues at the C-terminal end of the third cytoplasmic loop results in a constitutive, agonist-independent activation of the receptor (also

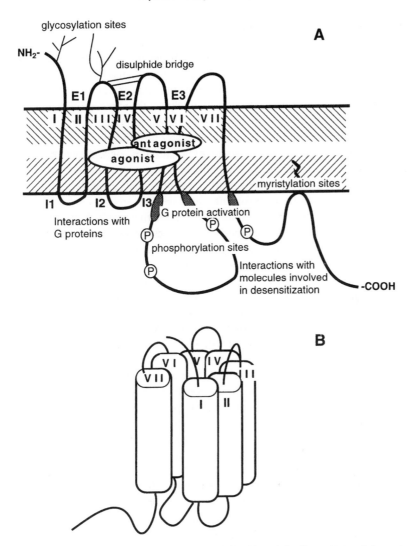

Fig. 4. The main features of the structure–activity relationships of the G-protein coupled receptors are presented in (A). The N-terminus bearing the glycosylation sites is extracellular whereas the membrane-anchored C-terminus is intracellular. The ligand-binding residues are deeply embedded at the inner side of the seven transmembrane α-helices (numbered from I to VII). The extracellular loops (E1, E2 and E3) bear glycosylation sites and disulphide bridges and the intracellular loops (I1, I2 and I3) are devoted to the interactions with the G proteins and other regulatory components (see text for details). The overall conformation of the G-protein coupled receptor is presented in (B), the cylinders schematising the transmembrane α-helices being folded in an anticlockwise manner to form a kind of pore in the plasma membrane.

named "intrinsic" receptor activity). Although changes in receptor activity have since been obtained for a few residues located outside the C-terminal part of the third cytoplasmic loop, these data pointed to this latter region as the key determinant to transmit the conformational changes of the agonist binding into another kind of conformation on the G protein α subunit (Samama et al., 1993; Perez et al., 1996). The observation of the constitutively activated receptors had a strong impact on the pharmacology and biochemistry of the G-protein coupled receptors (Lefkowitz et al., 1993). Firstly, these data underlined the allosteric character of receptors oscillating between different functional states, since they demonstrated that these latter can be obtained without the informative presence of ligands. Secondly, they promoted a better view of what was previously called receptor activity and of how agonists and antagonists interfered with it. Generally, when the receptor is empty, the inactive state predominates, although variably from one receptor subtype to another. Agonists are ligands which tend to displace the resting equilibrium toward the active conformation whereas antagonists do the opposite. Thirdly, taking into account the "intrinsic" activity of receptor facilitated the understanding the effects of certain drugs which tend to lower the basal activity of receptors, acting as so-called "inverse agonists". Whereas true antagonists do not modify the constitutive activity of receptors, inverse agonists stabilize the inactive state of the receptor leading to suppress any kind of receptor-dependent activity in cells. This phenomenon has obviously important therapeutic implications which only begin to be systematically explored.

The role of receptor as exchange factor for the α subunit of the G protein implicates an alteration of the structural interactions between these two proteins. Expectedly, the cytoplasmic regions (the cytoplasmic loops and the C-terminus) of the receptor play a keyrole in this function and three sets of observations support this contention (Fig. 4a). Firstly, mutagenesis experiments have identified the junctions between the third cytoplasmic loop and the adjacent transmembrane segments, as well as the C-terminal end of the receptor, as essential for the efficiency of G protein activation. It has been suggested that these protein stretches are able to form amphipathic α helices, similar to that of mastoparan, a wasp venom component able to efficiently activating G proteins. The current hypothesis states that, in the activated state, the membrane–cytoplasm junction of the sixth transmembrane segment is displaced toward the cytoplasm and exposes a short amino acid motif responsible for the opening of the nucleotide site on the α subunit of the G protein. It triggers the GDP–GTP exchange, that is the basic role of the receptor as a transduction device. The C-terminal end of the receptor (which forms a fourth cytoplasmic loop via its anchorage in the plasma membrane) is also required for the full range of receptor activity to be transmitted to G proteins.

Secondly, the use of careful mutagenesis and of peptides corresponding to the sequences of each of the cytoplasmic loops clearly showed that not only the third cytoplasmic loop is important for G protein activation (Dalman and Neubig, 1991; Arora et al., 1995; Gudermann et al., 1996). In this respect, mutagenesis approaches appeared generally inefficient since mutations in the two first cytoplasmic loops promoted a deleterious instability of receptor structure that generally impaired the proper folding and membrane insertion of the receptor (von Heijne and Manoil, 1990). Instead, all the loops are required for the purpose of G protein activation, even if the two first cytoplasmic loops are probably more important in the stabilization or in the positioning of the heterotrimeric G protein toward the receptor than in nucleotide exchange itself.

Thirdly, "natural" mutants of the cytoplasmic regions of the receptors exist in the form of splicing isoforms which change the length and the sequence of the third cytoplasmic loops and the C-terminus. Generally, these receptor isoforms exhibit different kinds of interactions with G proteins. It provides a single receptor subtype with the ability to couple several types of G proteins or with more or less potent activity. As demonstrative examples, the longest isoform of the D2 dopamine receptor (differing from the short one by the addition of twenty nine amino acids in the third cytoplasmic loop) could preferentially interact with Gi2 protein, whereas the short isoform would be more potent in cells were Gi2 is absent (Montmayeur et al., 1993; Guiramand et al., 1995; see also Valdenaire and Vernier, 1997). Also, some variants of the C-terminus of the prostaglandin E3 receptors activate adenylyl cyclase via the Gs protein whereas another variant activates the phosphoinositide turnover via a pertussis-toxin sensitive G protein (Hasegawa et al., 1996). Nevertheless, the determinants of either potency of G protein activation or of the specificity of receptor–G protein interaction have not been precisely identified. These properties are likely to be loosely defined on a structural basis, since most of the receptors are able to activate a wide range of G proteins in many different types of cells. Thus, the specificity and efficiency of coupling probably depend on the nature of the receptor and G proteins which are promiscuous in the membrane. Further understanding will soon come from more detailed structural approaches.

Since the G protein is the central component of the transduction process, most of our current view of the mechanisms of receptor–G protein interactions relies on the knowledge of G protein structure. The general structures of the α and $\beta\gamma$ subunits of the G proteins have been recently obtained both dissociated from and associated to each other (see Clapham et al., 1996; Neer and Smith, 1996 for reviews). This achievement brought important information on the mechanisms of G protein activation and interactions with receptors and effectors. The α subunit can be broadly divided into two parts, the globular GTPase domain and a helical regulatory domain. The GTPase domain appears very similar to that of any other GTPase of other families such as Ras and the other low molecular weight

G proteins (Rab, Rho, Ran, ...) or such as the translation elongation and initiation factors (Bourne et al., 1990). It has been shown that the N-terminus of the α subunit is required for its interaction with the $\beta\gamma$ subunits and other evidences point to the interaction of its C- terminal end with receptors (reviewed in Neer, 1995). Nevertheless, this C-terminal stretch is not sufficient for the receptor to promote the GDP–GTP exchange at the nucleotide site of the Gα protein and switch regions are also involved. The α subunit may also be associated to the plasma membrane by lipid modifications which vary from one class of G protein to another and may be the site of specific regulation (Neer, 1995). The core of the GTPase domain of the Gα protein is asymmetrically located in front of the β subunit, the exchange of GDP for GTP promoting a dramatic orientation change of the switch regions. This represents the probable mechanism of the heterotrimer dissociation which, in turn, triggers the modulation of effector activity.

The β subunit is characterized by short peptide repeats flanked by tryptophane and aspartate residues (the WD repeats) which form a highly symmetric ring structure similar to the blades of a propeller (Neer and Smith, 1996). In sharp contrast, the γ subunit is tightly bound and extended along the β subunit surface. Its main role could be to stabilize the β subunit structure and to anchor it into the plasma membrane since the γ subunit is prenylated at its C-terminus. The overall conformation of the α subunit depends on its association with the $\beta\gamma$ subunits, whereas this latter is strikingly similar whether it is coupled or not to the α subunit. Thus, the $\beta\gamma$ subunits are generally considered as a rigid scaffold used to stabilize several kinds of bound molecules. In this respect, it is worth mentioning that the propeller shape of the β subunit is not only used by the α subunit. It is also well suited to bind a special protein domain named "pleckstrin homology domains" (PH domains), present in many protein partners of the $\beta\gamma$ subunits (G-protein coupled receptor kinases, protein kinase Cμ, dynamin, ...). The role of the PH domain could be to localize effector proteins close to the membrane, in response to a given signal, such as the one promoted by G-protein coupled receptors (Shaw, 1996).

It becomes more and more apparent that $\beta\gamma$ subunits are playing an increasing number of roles in a broad range of interactions with intracellular effectors and regulatory proteins. For example, it is now clearly demonstrated that $\beta\gamma$ subunits are required to modulate one of the MAP kinase pathways, the signalling cascade which is commonly activated by tyrosine-kinase receptors and the Ras proteins. Although the complete set of intermediary proteins between $\beta\gamma$ and Ras is possibly not yet recognized, the membrane localization of Ras and the different MEK proteins are crucial for ERK activation. It involves the phosphorylation of the Shc protein on a tyrosine residue and the formation of the Shc-Grb2 complex required for Ras activation. The $\beta\gamma$ subunits provide the substrate for the membrane recruitment of these signalling proteins (van Biesen et al., 1995).

To conclude, G protein activation by receptors represents a succession of conditions to be fulfilled to ensure specificity and directionality in the modulation of effector activity. The activated state of the receptor is the condition to promote relocalization and interaction between the signalling molecules. The interactions between receptors, G protein and effectors are mutually exclusive and strictly depend on the state of G protein activation. This modular organization of transduction proteins conveys a high degree of temporal and spatial accuracy which characterize, despite their large diversity, the transmission of messages across the plasma membrane.

3. How cells modulate signal transmission by G-protein coupled receptors: from biosynthesis to regulation at the plasma membrane

Since membrane receptors are the molecules which receive and transmit extracellular signals, they are also the first place where cells can regulate their sensitivity and response to these signals. Regulation of receptor functions deserved a lot of attention these last years, since they are involved in phenomenons such as the vanishing effects of many drugs, tolerance to substances of abuse and withdrawal symptoms. It concerns mechanisms of receptor uncoupling, endocytosis from the plasma membrane and recycling which comprise the short-term desensitization process and the longer-term down-regulation. Their consequence is to turn-off or to attenuate the signalling events triggered by the interaction of extracellular messengers with the membrane receptors. Recent studies point to a general scheme of receptor regulation by sequential steps of phosphorylations and membrane retrieval of the receptors (Freedman and Lefkowitz, 1996). However, given the small number of receptor classes in which these phenomena have been studied in detail, the proposed mechanisms cannot be taken as granted for all the G-protein coupled receptors.

3.1. Desensitization and down-regulation of G-protein coupled receptors and G proteins

The main steps of G-protein coupled receptors desensitization have been described at best for the β-adrenoreceptors and rhodopsin (Freedman and Lefkowitz, 1996). As soon the receptor is uncoupled from the dissociated heterotrimeric G protein, its cytoplasmic regions become phosphorylated by two types of serine/threonine kinases. The second messenger-dependent kinases (protein kinase A (PKA) and protein kinase C (PKC)) are able to potently phosphorylate receptors at specific sites (Fig. 5). These kinases can be activated either by the receptor itself which will undergo to the desensitization process or by neighbour receptors of different specificity (heterologous desensitization). The physiological relevance of heterologous desensitization is not clear at present. The second type of kinases which regulates the interactions of receptors with other intracellular components is named G-protein coupled receptor kinases (GRK). They come into play only when receptors are dissociated from the G protein and they are specifically phosphorylating activated receptors, that is agonist-occupied receptors. Accordingly, they are only implicated in homologous desensitization processes, but not in the regulation of another neighbouring receptor. They form a protein family comprising six known members which differ by the way they are associated to the plasma membrane (reviewed in Ferguson et al., 1996).

Many classes of G-protein coupled receptors have been shown to be regulated by these kinases. For example, the β_2-adrenergic receptor may be desensitized by five of the GRK which were also shown to promote the desensitization of rhodopsin (Inglese et al., 1993). When receptors are in the resting state, GRK are localized in the cytosol. Their necessary translocation to the plasma membrane is triggered by receptor activation by a variety of different mechanisms. Alternatively, GRK (GRK2 and 3) may be farnesylated at the C-terminus (GRK1) and localized by this means at the plasma membrane or targeted at the right place via interaction with the isoprenylated $\beta\gamma$ dimer of G proteins. In this latter case, the PH domain of the GRK binds to the WD domains of the β subunit. Finally, other GRK are associated to the membrane by either palmitoylation (GRK4, GRK6) or a stretch of basic residues (GRK5). Receptor palmitoylation/myristylation which anchors the C-terminal end of the receptor the to the plasma membrane is probably involved in the mechanisms underlying the specific effect of GRK on activated receptors. The hypothesis will be that the active state of the receptor triggers the depalmitoylation of the receptor, rendering it more accessible to phosphorylation.

The full desensitization of the receptor requires the contribution of additional components such as the arrestin proteins. Only two arrestins, related to the visual rod and cone arrestins, have been identified so far. Here again, arrestins need to be translocated to the plasma membrane where they preferentially bind to phosphorylated receptors, making up the uncoupling from the G proteins. The binding of

---

Fig. 5. This general model of receptor desensitization is based on the data obtained for the β adrenoreceptors, rhodopsine and related receptors. (1) Agonist occupation of the receptor promoted its phosphorylation by second-messenger dependent kinases (PKA is this case). (2) G-protein coupled receptor kinases (GRK) are then recruited at the plasma membrane by their interactions with $\beta\gamma$ subunits, and pursue receptor phosphorylation and uncoupling from the α subunit of the G protein. Other proteins such as RGS (regulators of G protein signalling) also modulates the response promotes by the activated receptors. (3) The phosphorylated receptor then binds arrestin which is thought to interact with the endocytotis machinery illustrated here by the presence of the coat proteins of the forming vesicle. A role of the heterotrimeric G protein in this process is possible but still unclear.

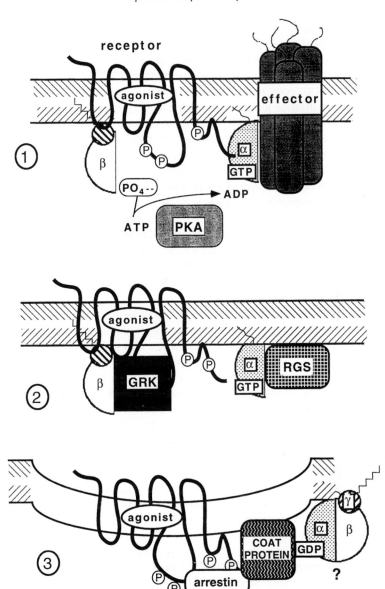

arrestin to the receptor could contribute to the initiation of the receptor sequestration (Ferguson et al., 1996). In this process, the role of arrestin will be analogous to that of an adapter protein which favours the association of the receptor with coat proteins (such as clathrin) of the endocytotic vesicle (Fig. 5). In addition, it is now clear that receptor phosphorylation by GRK and other kinases is not required for internalization to occur, although it renders the process faster and more powerful. In contrast, the presence of arrestin is mandatory, as shown by the fact that receptor sequestration can be achieved normally in the absence of receptor phosphorylation by overexpressing arrestins (Ferguson et al., 1996). It has been suggested that the receptor internalization could be involved in a resensitization and recycling process since only a low proportion of phosphorylated receptors as well as the high phosphatase activity were found to be associated with the intracellular sequestration compartment. Contrasting with what was previously suggested, only a small part of the internalized receptors are directed to the lysosomes to be degraded. The nature of the vesicles which transport endocytosed receptors is still elusive. For instance, both clathrin-coated vesicles and caveolæ were reported to harbour the sequestered β_2-adrenergic receptors but arrestins have been shown to bind only the clathrin coat of vesicles

To summarize, agonist activation of the receptor results in the translocation of the GRK to the membrane, essentially by binding to the free $\beta\gamma$ subunits. Then, receptor phosphorylation increases affinity for arrestins which, by interaction with coat proteins, direct the receptor to the endocytotic vesicles. Internalized receptors are then dephosphorylated, and can be sent back to the plasma membrane. The mechanisms underlying this last transport are not known, but, by analogy with the recycling of other proteins, it may involve the sorting of the dephosphorylated receptor from the late endosomes to the trans-Golgi network and then to the plasma membrane, although a more direct pathway to the plasma membrane may be envisaged.

Beside the mechanisms of receptor desensitization and down-regulation, which tend to prevent receptor–G protein interactions, these latter also undergo a regulation in order to turn-off quickly their activity. As already mentioned, the intrinsic GTPase activity of the α subunit of heterotrimeric G proteins is low. In particular it cannot achieve the fast inactivation required by many signalling processes. Recently, a new family of proteins has been discovered and named RGS (for Regulatory of G protein Stimulation). They have been isolated on the basis of genetic screens for mutant animals (yeast, *Aspergillus*, nematodes and flies) which failed to desensitized to different kind of membrane transmission. Mammalian counterpart of these regulatory proteins have also been isolated and no less than sixteen different RGS are already known. On a biochemical basis, RGS bind to the GTP-associated form of the G protein and speed up the GTP hydrolysis, without affecting GDP-GTP exchange or affinity G protein for the nucleotide. Thus, they act as a complement to the intrinsic GTPase-activating component of the Gα protein to render the GTPase activity compatible with the physiological requirement of fast signal transmission at the plasma membrane. There is also some possibility, based on weak sequence similarities, that RGS themselves are GAP proteins. The specificity of each of the recently isolated RGS toward a given G protein class seems to be poor in our present stage of knowledge (see Koelle, 1997; Dohlman and Thorner, 1997). However, two features point to differential activity or functions for the various RGS. Firstly, some of them have additional sequence domain which could bind to other protein partners, such as phosphatases in the case of the *fused* gene product in *Drosophila*. Secondly, most of RGS have a restricted pattern of distribution, also suggesting that their expression is spatially organized in the whole organism, and probably also during embryogenesis.

3.2. The biosynthesis and intracellular transport of G-protein coupled receptors

The mechanisms of receptor biosynthesis, of their integration into the intracellular membranes, of their transport to – or from – the plasma membrane has retained much less attention than their pharmacology and structure–activity relationships. Nevertheless, the recent advances of cell biology have now provided studies on G-protein coupled receptors with new techniques of receptor visualisation. In particular, the possibility to fuse protein "tags" with the receptor sequences allowed to use immunocytochemical techniques to accurately follow the receptor behaviour inside cells with confocal, video or electron microscopes.

As all the integral membrane proteins, the seven transmembrane G-protein coupled receptors are co-translationally integrated in the membrane of the endoplasmic reticulum. A large majority of G-protein coupled receptors do not have a signal sequence, the structure used by the "signal recognition particle" (SRP) to promote the move of the nascent protein chain through the pore of the translocation complex in the membrane of the endoplasmic reticulum (reviewed in Schatz and Dobberstein, 1996). Nevertheless, the hydrophobic stretches of the G-protein coupled receptors (which will become the transmembrane segments after membrane translocation) are flanked by polar residues which define signal-anchor sequences used to determine the proper orientation and insertion of the protein in the lipid membrane (Audigier et al., 1987; von Heijne and Manoil, 1990). The alternance of hydrophobic stretches (signal-anchor) and charged residues (stop transfer) seems to be the structural information required for membrane insertion of the G-protein coupled receptors (High and Dobberstein, 1992). During this membrane translocation process, the first hydrophobic segment of the receptor is inserted alone into the reticulum membrane, and then the six other segments are inserted two by two through the translocation pore complex (Audigier et al., 1987). The final oligomerization of the receptor probably depends on the contact of the lipophilic part of the α-helices and on the interaction of polar residues and the formation of ionic and hydrogen bonds at their water-exposed faces (Popot and Engelman, 1990). In some cases, it is also possible that the proper fold of the receptor required the presence of a specific chaperone, as recently demonstrated for the mammalian red opsin (Ferreira et al., 1997).

Then, the receptor is glycosylated at N- and O-sites in the lumen of the endoplasmic reticulum, as any membrane protein (reviewed in Abeijon and Hirschberg, 1992). The precise role of these glycosylation events is not clear, since they are not required for the receptor to bind ligands and to perform correct transduction process. However, impairment of the receptor glycosylation resulted in a decreased number of receptor at the cell surface in the case of the β_2-adrenoreceptor and rhodopsin (Rands et al., 1990). Probably, several other post-translational modifications occur in the endoplasmic reticulum and Golgi apparatus, but most of them have not been studied yet. Among them, it may be reminded that the myristoylation of the C-terminus of the receptor protein (which probably takes place in the Golgi complex) is strictly required for the receptor to efficiently transduce extracellular signals.

The final membrane sorting of the receptor is probably achieved in the trans-Golgi network (TGN) where the decision of transporting it to different domain of the plasma membrane takes place (Fig. 6). This step is of special importance since it will govern the differentiation of microdomains of the plasma membrane specialized for receiving extracellular messages and it will spatially organize cell communication pathways in a metazoan organism (Lisanti et al., 1994). The example of a polarized cell such as neurones or epithelial cells is particularly striking, since the receptors have to be localized either to the dendritic of the axonal compartment (in the case of neurones) or to the basolateral or apical domains (in the case of epithelial cells). Although the mechanisms of this specific sorting are still essentially mysterious, two hypothesis may account for the receptor targeting at their final location.

The first one states that the type and length of the transmembrane α-helices governs the nature of the membrane where the protein will be finally sorted out (Munro, 1991; Pelham and Munro, 1993). Here, there is a kind of mutual recognition of the integral protein by the membrane which is compatible for it. The existence of a differential lipid composition of microdomains in the intracellular and plasma membrane makes this hypothesis attractive. The second hypothesis stipulates that the cytoplasmic regions of the membrane proteins act as targeting signals which may be recognized by specific protein acting as scaffold for the proper location of the proteins at the plasma membrane and for their association with the other proteins of the transduction modules. Examples of such scaffolding proteins are PDZ domain proteins. These proteins have no functional effect on their own but they help to localize properly at the membrane functional pro-

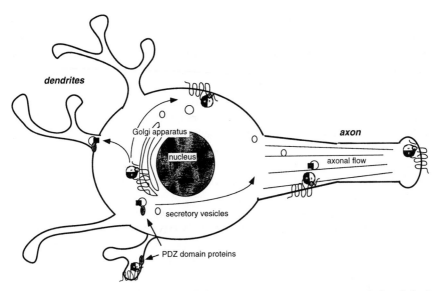

Fig. 6. A schematic view of the biosynthesis and sorting of the G-protein coupled receptors in polarized cells such as neurones. In these cells, the information flows goes from dendrites to axons. Localization of receptors associated to their coupled G proteins in either dendrites or neurones are of utmost importance for the way these cells will respond to the neurotransmitter. The sorting of the receptors is probably achieved when they go out of the Golgi at the end of the trans-Golgi network and could implicate G protein themselves as well as specific localization proteins such as PDZ proteins.

teins such as receptors and to organize the interactions with effector partners. The major role of PDZ proteins is for example spectacularly illustrated by the *InaD* mutation in *Drosophila* which alters the distribution of rhodopsin and of the coupled phospholipase C and calcium channel in the photoreceptor cells of the eye (Tsunoda et al., 1997). It is now a general observation that many proteins exist just to direct the interactions of components of signalling pathways, underlining the importance of subcellular localization for molecular functions (reviewed in Pawson and Scott, 1997).

The regulation of receptor sorting in the secretory pathway could also implicate interactions with heterotrimeric G proteins. Activation of the Go/Gi proteins or of Gs, both of which are abundantly present in the Golgi and the TGN has an inhibitory effect on the formation of secretory vesicles and on the transport of secreted proteins (Leyte et al., 1992). In contrast, the blockade of Go/Gi proteins by the pertussis toxin stimulates vesicle budding in the TGN. In addition, it has been suggested that several types of heterotrimeric G proteins are able to regulate the specific localization of membrane proteins in polarized epithelial cells (Pimplikar and Simons, 1993).

In neurones, receptors are found not only on the cell soma but also at specific places in the membrane such as dendritic shafts, in the presynaptic button or all along the axons (Fig. 6). Although the possibility of a tight regulation of these transport mechanisms exists, neither their existence nor a firm hypothesis can be established up to now. In particular, the link between the processes of desensitization and down-regulation and those of membrane transport is not clear although these phenomenons are just particular cases of more general mechanisms of regulation of receptor localization and activity.

4. The molecular diversity of transmission modules at the plasma membrane

As already stated, the combinations of the three G protein subunits with receptors and effectors possibly leads to an eminently large diversity of membrane signalling modules (Fig. 2). The question of the experimental approaches which have to be designed to deal with such a multiplicity of interacting molecules is a difficult challenge for modern cell biology. It is very similar to the general problem of "complexity" (i.e. the combination of many different components that cannot be simplified to a few interchangeable elements) commonly encountered in all fields of biology. Resolving this issue is currently out of reach, but a few pitfalls and some directions may be indicated.

At first glance, the specificity of the G-protein coupled receptors with their patterns does not seem to be striking, since most of the receptors are able to activate more than one type of G proteins. Most of the time, when receptors are transfected in heterologous cells, significant activation of different signalling pathways can be obtained, showing than G proteins and receptors are easily promiscuous and able to interact with each other. In these case, the response elicited in cells will just depends on the nature of the effectors which are present or recruited at the membrane. Of course, this kind of approaches, although very widely used may miss some crucial aspects of the functions of the transduction module. Receptor transfections remain a useful tool to study many aspects of structural biochemistry and molecular pharmacology of G-protein coupled receptor, but certainly not to predict the responses a given receptor may elicit in a natural system.

On the other hand, it is equally wrong to state that everything is possible when speaking about receptor–G protein interactions. First of all, the molecular diversity found in natural situations is not as large as it could theoretically be. There are preferential association of receptors with G proteins and for the activation of effectors by the G protein subunits. In addition, some incompatibilities exist in the composition of heterotrimers. For instance, it has been shown that a trimer cannot be formed between β_2 and γ_1 subunits and many preferential interaction between α, β and γ have been described (see Neer, 1995 and Gudermann et al., 1996 as reviews). Similarly, α_1 adrenoreceptors are able to interact with G_q proteins only, and muscarinic M_2 receptors, α_2 adrenoreceptors, or dopamine D_2 receptors only activate G_i/G_o proteins.

A high degree of specificity is probably more manifest for the modulation of effectors by G proteins. All the subtypes of adenylyl cyclase are activated by $G\alpha_s$, but only type II and IV are activated by $\beta\gamma$ which also potentiate the effect of $G\alpha_s$. In contrast, $\beta\gamma$ has no effect on the type III adenylyl cyclase and it inhibits the neuronal type I cyclase (Sunahara et al., 1996). It is thus reasonable to propose that, in natural situations, the preferential association of receptor classes with groups of heterotrimeric G proteins are the rule, a full compliance of interactions being exceptional (Gudermann et al., 1996).

Which kind of picture could we now propose to account for the required specificity in membrane signalling with this paradoxical multiplicity and diversity in the molecules making the transmission modules? An important aspect of the question is to figure out that this molecular diversity and multiplicity takes place inside large organisms, to meet many different requirements throughout life. Thus, in a complex vertebrate organism, the release and action of extracellular transmitters (hormones, chemoattractants, neurotransmitters, . . .) have restricted temporal and spatial organizations which vary from one system to another. This is apparent both at the level of the whole organism and at that of the cell. Demands for receptor functioning are not the same for fast synaptic transmission leading to ion channel activation and for the slow signalling displayed by many neuromodulatory or hormonal systems.

A large degree of the molecular diversity of receptors and intracellular partners account for this necessity of anatomical organization. It is demonstratively observed in the nervous system where most of the different subtypes of neurotransmitter receptors are localized in different areas, in different neurones, in different types of cells. The prevailing status is that the various subtypes of a given receptor class are localized in a nonoverlapping fashion both in the brain and in peripheral tissues. Taking the β-adrenoreceptors as an example, β_1 receptors are found in the anterior olfactory nuclei, in the cerebral cortex, striatum, in deep cerebellar nuclei, whereas β_2 receptors are found in the olfactory bulb, and the piriform and cerebellar cortex, and the β_3 receptor is not present in the brain (Nicholas et al., 1996). Therefore, these different receptor subtypes cannot be considered as redundant, even if they are doing very similar effects in cells, since they are not localized in the same cells. Instead, they carry out the necessary transduction of the extracellular transmitter or hormones in different target cells. To a lesser extent, the different isoforms of a the transduction protein such as the α subunit of G proteins could be expressed in a defined anatomical manner (G_{olf} versus G_s, transducin versus G_i for example), as well as many of the variants of effector proteins. Thus, the anatomical organization of the transmitter receptors and of their coupled proteins accounts for the largest fraction of the specificity of information transmission to fulfill the whole range of regulatory functions.

At the level of cells, the question should be tackled differently. Indeed, it is generally observed that, for a given cell, the range of responses to several different transmitters is not extremely large. Cells have to respond to numerous messages, but the number of responses elicited by these stimuli are rather limited: Extracellular signals converge towards a few adaptive responses in each cell type. Accordingly, the number of proteins which compose intracellular signalling pathways is much smaller than that of transmitters and receptors. Striving is still required to better know the natural composition of receptor modules in a given cell. A fundamental piece of the functional specificity of signalling process is provided by the sorting mechanisms which specifically address receptors and their interacting partners to membrane compartments. It has already been stressed that specialized domains of the plasma membrane exist, such as caveolæ or post-synaptic densities, where signalling components are concentrated (Lisanti et al., 1994). The final specificity of the cellular responses to different signals depends on the biochemical properties and localization of the receptors molecules and on the nature of the very few effectors which ultimately determine phenotypic responses. In other words, if the incoming messages and the final responses need to be highly specific for a cell type, the molecules which carry the signals between them do not.

An other way to reach a high degree of specificity in cell signalling is the temporal coherence of the many messages which reach the cell. Attention has been given these past years to what is called "coincident detection" (Bourne and Nicoll, 1993). It means that the temporal order of receptor activation and the duration of the elicited responses introduce a large amount of variability and specificity to rather stereotyped molecular interactions. It also provides a means to make signals coming out of the background noise. As an example, the activity of the type II adenylyl cyclase is additively potentiated by α_s and $\beta\gamma$ subunits, when they are acting together. Therefore, a weak stimulation of a receptor activating Gs which will not be able to activate adenylyl cyclase may become efficient if an other type of receptor is able to release a small amount of $\beta\gamma$ subunits. Regarding reaction timing in signal transmission, the role of G proteins is clearly to impose a temporal pattern to biological events through the kinetics of cycles, its nucleotide exchange and hydrolysis. In this respect, the consequence of receptor activation is really the result of the receptor–G protein–effector interactions since at each of these protein interfaces regulation may occur. The recent discovery of RGS proteins (see above) which affect the rate of GTP hydrolysis by the α subunits is important in this respect although their place in G-protein coupled receptors transmission still needs to be evaluated. In addition, the various receptors subtypes may also exhibit variable rates of desensitization and internalization, as it is the case for β-adrenoreceptor subtypes.

The state of functional differentiation of the cell will also strongly affect the nature of the final responses. This is well demonstrated in the prolactin cell of the anterior pituitary. In a male or non-pregnant female, hormone release is efficiently inhibited by the tonic effect of low concentration of dopamine. This transmitter, acting via D_2 receptors and $G\alpha_i$ proteins, promotes the hyperpolarization of the cell membrane by enhancing the permeability of potassium channels (Lledo et al., 1992). But during lactation, many of the prolactin secreting cells display a completely different behaviour with a low resting membrane potential and accordingly high and fluctuating levels of intracellular calcium, leading to a high level of spontaneous hormone release. In these cells, the D_2 receptors trigger $G\alpha_o$ activation and calcium channel blockade promoting hormone release inhibition. Moreover, in lactating females, other prolactin cells with high membrane potentials can release prolactin only upon TRH receptor activation, and become sensitive to dopamine only in this condition (Lledo et al., 1992, 1994). Therefore, most of the apparent "pleiotropy" of G-protein coupled receptors could be resolved when the various functional states of the cells and spatial distribution of the signalling systems are taken into account.

To summarize, specificity in cell signalling is essentially provided by cell differentiation (which governs the nature of the interacting components of G-protein coupled receptors signalling modules), by the specific localization and compartmentalization of molecular components in the cells, by their stoichiometry and the precise timing of the recruitment of the interacting partners making the so-called signalling pathways.

5. The generation of receptor multiplicity in vertebrates: an evolutionary approach

The combination of sophisticated techniques of molecular and cell biology which has been applied to the study of receptor activity this five past years, although providing a better account of the receptor structure–activity relationships, did not say much on the physiological role of the receptors in a whole metazoan organism. It is probable that molecular activities can be simplified to coordinated interactions of proteins belonging to a given functional module, which exists only in the context of a particular cell. However, these cells are assembled in networks and organs, each of which contribute to the physiological behaviour of the organism and to its adaptation to many different situations and events throughout life. This organization especially underscores the importance of the differentiation process and of the spatial arrangement of cells in the organization of an organism. Regulations, that is the collection of functions which ensure the adapted changes of other functions inside a whole organism, provide another level of general organization.

When one considers the activation or inhibition of one single receptor molecule, it is actually impossible to infer the consequence for an organism. It sets the limits of the biochemical and cellular approach to general physiology (Humphrey et al., 1993; Vanhoutte et al., 1996). This epistemological difficulty can be overcome by a genetic approach the goal of which is precisely to link a given genotype (the molecule) to one or several phenotypes (the physiological consequence in the organism). However, experimental genetics is applicable to only a small number of organisms up to now, and phenotypes are far from being known for the majority of the proteins. This is particularly the case for most of the G protein receptors. In addition, it is difficult to fully understand the role of a given molecule without taking into account the historical dimension which has shaped both molecule and organisms, the dimension of biological evolution*.

Therefore, understanding how the genetic multiplicity of G-protein coupled receptors has been generated, and which is its functional counterpart, requires to take into account the strong and parsimonious logic of evolution, as it is generally the case for molecular systems (Cardinaud et al., 1997; Vernier et al., 1995; Zuckerkandl, 1994; Donnelly et al., 1994).

5.1. Generalities about molecular evolution

The fact that receptors such as G-protein coupled receptors form large families is the hallmark of the evolutionary origin of the observed molecular multiplicity. It is necessarily the outcome of random processes of molecular diversification, of some degree of contingency and of a part of real functional and adaptive necessity. These multiple and different forms of resembling proteins are the result of many steps of gene duplications that occurred at different times of species evolution. Whatever the mechanisms of these gene duplications would be, the large number of point mutations which occur after gene duplication drives a sequence drift leading to differentiate the two duplicated sister genes (paralogous genes). Most often, this genetic drift leads to inactivation of one copy of the redundant sister genes which are thus transformed into pseudo genes. To be conserved, duplicated genes have to escape this inactivation, generally by acquiring new functions selected as essentially nonredundant and associated with identifiable changes in the physiology of the corresponding species. Thus, the duplicated paralogous genes are then conserved in all the species which will derive from the ancestral population where the duplication occurred.

A second kind of molecular diversity is provided by the disparity between animal species. Indeed, when a significant group of interbreeding animals be-

---

* This is best stated by the famous Dobzhansky aphorism: "Nothing in biology makes sense except in the light of evolution".

comes separated from its ancestral population for any reason, the two lineages will rapidly differ from each other, from a genetic point of view. This process is called "speciation" and the genes expressed in the two separated population will also soon become sufficiently different to be distinguished as "orthologous genes" (i.e. same genes in different species). As far as receptors are concerned, gene duplications give rise to classes and subtypes of receptors whereas speciation provides only species homologues (orthologous receptors). Receptor classes and subtypes necessarily diverged before a speciation event occurred and are thus found in all the species which derived from the ancestor where the gene duplications took place. Given the regular rate of the neutral genetic drift, orthologous genes will always display a higher degree of sequence similarity than paralogous subtypes or classes of receptors. This property can be exploited to use sequence alignments and comparisons to build receptor classifications, which should represents the true phylogeny of the receptors (Felsenstein, 1988; Vernier et al., 1995).

5.2. *The evolution of bioamine receptors in vertebrates*

When molecular phylogenetic classifications have been applied to the G-protein coupled receptors specific for bioamines, two main conclusions have been reached. Firstly, all the receptor classes previously identified on a pharmacological basis are also recognized by the molecular phylogenies. In addition, it has been surprisingly observed that receptor classes which bind a given neurotransmitter are not more closely related to each other than they are to other classes of receptors (for example α_2 adrenoreceptors are not more closely related to α_1 adrenoreceptors than they are to dopamine D_2 or serotonin $5HT_2$ receptors; Fig. 7). It means that these receptor classes which are specific of the same neurotransmitter do not share a direct common ancestor, and that they acquired independently and convergently the ability to bind this transmitter. Secondly, these analysis of the molecular phylogenies of bioamine receptors also revealed that many sequences exhibited a variable rate of apparent sequence divergence (Fig. 7). This parameter is the reflection of the random rate of mutations which strikes similarly all the genes of a given species, on the one hand, and, on the other hand, of the structural constraints which allow only a limited number of nearly neutral mutations to be conserved. The others are eliminated as deleterious or incompatible with receptor function in the cells and organisms in which they are expressed. Despite this various rate of divergence, it remains clear enough that most, if not all the bioamine receptor subclasses found in vertebrates are generated during the evolution of this phylum.

A closer look at the constraints which shaped structure and functions of the G-protein coupled receptors will highlight the physiological "raison d'être" of their molecular diversity. To be able to do that properly, it is necessary to have isolated the different receptor classes and subtypes in as much as possible of an-

188 P. Vernier

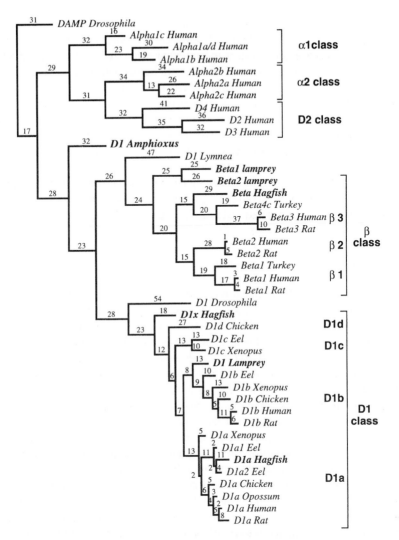

Fig. 7. A phylogenetical tree of catecholamine receptors with special emphasis on dopamine D_1 and β-adrenoreceptors. This tree has been obtained by the parsimony analysis (by using the PAUP 3.1.1 parsimony software) performed on alignments of the receptor sequences obtained in species representative of many groups of vertebrates. It is rooted to the *Drosophila* DAMB receptor which is more closely related to α_1, α_2 and to D_2-like receptors than to D_1 and β receptors. The sequence relationships respect the known vertebrate species phylogeny. For the α_1, α_2 and the D_2-like receptors, only human sequences encoding each of the corresponding subtypes have been selected. For D_1 and β receptor classes, the names of the vertebrates subtypes are indicated with brackets. The numbers indicated how many steps were required to derive each branch.

imal species which represent the main steps of vertebrate evolution. Up to now, such a study has been carried out only for the D_1 dopamine receptors and partly for the β adrenoreceptors. Results for the other classes of G-protein coupled receptors are partial but they confirm all the conclusions which can be drawn from the D_1 dopamine receptors. Basically, it seems that some of the receptor classes which are found in vertebrates have been either transformed from related classes found in invertebrates, or de novo created by gene duplications that occurred simultaneously with the emergence of vertebrates during evolution. For example, it is very likely that the vertebrate α_2 adrenoreceptor class arose by transformation of the invertebrate octopamine receptors whereas the β adrenoreceptor class specifically emerged in vertebrates from the duplication of an ancestor gene shared with the D_1 dopamine receptors. A descendant of this ancestral receptor is still found in cephalochordates (amphioxus), the closest group of living vertebrate relatives (Fig. 7; Cardinaud et al., 1998). This latter hypothesis is strengthened by the fact that D_1 and β receptors share many biochemical properties such as a very similar structure, an interaction with the same kind of G proteins (Gs) and the same mechanisms of receptor desensitization.

In the vertebrate phylum, two other steps of gene duplications took place to generate the present state of bioamine receptor multiplicity. The first of these events happened in the agnathe lineage (vertebrates without jaws), probably before, or simultaneously, to the emergence of hagfish, whereas the second one occurred with the output of jawed vertebrates. These two major steps of gene duplications have also appeared for many developmental genes expressed in the vertebrate nervous system. They may correspond to the dramatic genetic changes which underlie the remodelling in organogenesis which accompanied the emergence of vertebrates from ancestral chordates. Thus, our current hypothesis states that the new duplicated receptors were recruited subsequently to the changes promoted by the master genes of nervous system development and that they were used to achieve the increased demand in information processing.

Based on these data and observations a plausible scenario for the emergence and evolution of receptors in the vertebrate nervous system can be proposed. At the origin of vertebrates, several classes of bioamine receptors emerged by conservation of duplicated genes to fulfill the requirements of many demanding new functions. They were generated by the development of new sensory organs and of the derivatives of the neural crest such as the sympathetic peripheral nervous system or the partitioned heart. In all cases, the receptors developed many new characteristics as compared to their parental "chordate-but-non vertebrate" genes. However, these receptor characteristics became fixed and remained essentially unchanged in the whole vertebrate phylum. These characteristics include some determinants of the ligand binding which are delineated by a specific pharmacological profile, the coupling to a given set of G proteins, a specific localization in cells and some general features of receptor desensitization and regulation. When each of the two duplications occurred before or simultaneously to the emergence of gnathostomes (jawed vertebrates), one of duplicated genes often conserved most of the ancestral role, whereas the other copy was more free to acquire new functional peculiarities. These features became fixed as soon as they serve new adaptation requirements for the organisms which expressed them, and they were conserved as homologous characters in all the corresponding species. Among the mechanisms needed for receptor subtype diversification, the acquisition of a tissue-specific pattern of expression and of a cell-specific localization were of utmost importance for the conservation of the duplicated receptors.

Just to illustrate this contention, the D_{1A} subtype of dopamine receptor exhibits functional characteristics which are essentially the same in all species from hagfish to human, and which are therefore very likely to be close to those of the ancestral vertebrate D_1 receptor. In contrast, the D_{1B} receptor subtype shows some derived characters which are conserved in all the vertebrate species where it has been analyzed (a high affinity for dopamine as compared to the D_{1A} subtype and a significant degree of intrinsic activity). In addition, the different dopamine D_1 receptor subtypes are expressed in different areas of the central and peripheral nervous system, rendering these subtypes essentially nonredundant since dopamine will act independently on each its target cells which differentially expresses the receptor subtypes.

As a conclusion, it has to be stressed that even if the powerful tools of modern molecular biology have shed a bright light on how the G-protein coupled receptors are structured and how they act as informative components of signalling modules, their contribution to the function of organisms will need to take into account the results of genetics and comparative biology. The main characteristics of the nervous system organization in modern vertebrates have been elaborated by a few major steps of genetic remodelling. G-protein coupled receptors participated in these events as key regulatory components which allows cells to adapt to many new situations and environments. Elucidating the constraints that shaped the expression, structure and activity of these receptors will also lead to a wider and broader understanding of how their function may become maladaptive in so-called pathological states.

References

Abeijon, C. and C. Hirschberg, Trends Biochem. Sci. **17** (1992) 32.
Arora, K., A. Sakai, and K. Catt, J. Biol. Chem. **270** (1995) 22820.
Audigier, Y., M. Friedlander, and G. Blobel, Proc. Natl. Acad. Sci. USA **84** (1987) 5783.
Birnbaumer, L., J. Abramowitz, and A. Brown, Biochim. Biophys. Acta **1031** (1990) 163.
Bourne, H.R., D.A. Sanders, and F. McCormick, Nature **348** (1990) 125.

Bourne, H.R., D.A. Sanders and F. McCormick, Nature **349** (1991) 117.

Bourne, H.R. and R. Nicoll, Cell **72** (1993) 65.

Bouvier, M., S. Moffett, T. Loisel, B. Mouillac, T. Hebert, and P. Chidiac, Biochem. Soc. Trans. **23** (1995) 116.

Cardinaud, B., K.S. Sugamori, S. Coudouel, J.D. Vincent, H.B. Niznik, and P. Vernier, J. Biol. Chem. **272** (1997) 2778.

Cardinaud, B., J.M. Gibert, J.D. Vincent, H.B. Niznik, and P., Vernier, in: *Trends in Comparative Endocrinology and Neurobiology* (Ann. N.Y. Acad. Sci., 1998).

Chothia, C. and A. Finkelstein, Annu. Rev. Biochem. **59** (1990) 1007.

Clapham, D., J. Codina, and L. Birnbaumer, Nature **379** (1996) 297.

Dalman, H. and R. Neubig, J. Biol. Chem. **266** (1991) 11025.

Demchyshyn, L.L., K.S. Sugamori, F.J. Lee, S.A. Hamadanizadeh, and H.B. Niznik, J. Biol. Chem. **270** (1995) 4005.

Dohlman, H.G. and J. Thorner, J. Biol. Chem. **272** (1997) 3871.

Donnelly, D., J. Findlay, and T. Blundell, Receptors Channels **2** (1994) 61.

Ehrlich, P., Lancet **2** (1913) 445.

Felsenstein, J., Annu. Rev. Genet. **22** (1988) 521.

Ferguson, S.S., L.S. Barak, J. Zhang, and M.G. Caron, Can. J. Physiol. Pharmacol. **74** (1996) 1095.

Ferreira, P.A., T.A. Nakayama and G.H. Travis, Proc. Natl. Acad. Sci. USA **94** (1997) 1556.

Freedman, N.J. and R.J. Lefkowitz, Recent Prog. Horm. Res. **51** (1996) 319.

Grigorieff, N., T.A. Ceska, K.H. Downing, J.M. Baldwin, and R. Henderson, J. Mol. Biol. **259** (1996) 393.

Gudermann, T., F. Kalkbrenner, and G. Schultz, Annu. Rev. Pharmacol. Toxicol. **36** (1996) 429.

Guiramand, J., J.-P. Montmayeur, J. Ceraline, M. Bathia, and E. Borrelli, J. Biol. Chem. **270** (1995) 7354.

Hasegawa, H., M. Negishi, and A. Ichikawa, J. Biol. Chem. **271** (1996) 1857.

Henderson, R. and P. Unwin, Nature **257** (1975) 28.

Henderson, R., J. Baldwin, T. Ceska, F. Zemlin, E. Beckmann, and K. Downing, J. Mol. Biol. **213** (1990) 899.

Hibert, M.F., S. Trumpp-Kallmeyer, A. Bruinvels, and J. Hoflack, Mol. Pharmacol. **40** (1991) 8.

High, S. and B. Dobberstein, Curr. Opinion Cell Biol. **4** (1992) 581.

Hill-Eubanks, D., E. Burstein, T. Spalding, H. Brauner-Osborne, and M. Brann, J. Biol. Chem. **271** (1996) 3058.

Hoflack, J., S. Trumpp-Kallmeyer, and M.F. Hibert, Trends Pharmacol. Sci. **14** (1994) 7.

Inglese, J., N.J. Freedman, W.J. Koch, and R.J. Lefkowitz, J. Biol. Chem. **268** (1993) 23735.

Humphrey, P.P., P. Hartig, and D. Hoyer, Trends Pharmacol. Sci. **14** (1993) 233.

Javitch, J., D. Fu, J. Chen, and A. Karlin, Neuron **14** (1995) 825.

Kenakin, T.P., Trends Pharmacol. Sci. **4** (1983) 291.

Koelle, M.R., Curr. Opinion Cell Biol. **9** (1997) 143.

Langley, J.N., J. Physiol (London) **33** (1905) 374.

Lefkowitz, R.J., S. Cotecchia, P. Samama, and T. Costa, Trends Pharmacol. Sci. **14** (1993) 303.

Lisanti, M., P.E. Scherer, Z.L. Tang, and M. Sargiacomo, Trends Cell. Sci. **4** (1994) 231.

Leyte, A., F. Barr, R.H. Kehlenbach, and W.B. Huttner, EMBO J. **11** (1992) 4795.

Lledo, P.M., V. Homburger, J. Bockaert, and J.D. Vincent, Neuron **8** (1992) 1.

Lledo, P.-M., P. Vernier, L.A. Kukstas and J.-D. Vincent, in: H.B. Niznik (ed.), *Dopamine Receptors and Transporters* (Dekker, 1994) pp. 59.

Montmayeur, J.-P., J. Guiramand, and E. Borrelli, Mol. Endocrinol. **7** (1993) 161.

Munro, S., EMBO J. **10** (1991) 3577.

Neer, E.J., Cell **80** (1995) 249.

Neer, E.J. and T.F. Smith, Cell **84** (1996) 175.

Nicholas, A.P., T. Hokfelt, and V.A. Pieribone, Trends Pharmacol. Sci. **17** (1996) 245.

Pawson, T. and J.D. Scott, Science **278** (1997) 2075.

Pelham, H.R. and S. Munro, Cell **75** (1993) 603.

Perez, D., J. Hwa, R. Gaivin, M. Mathur, F. Brown, and R. Graham, Mol. Pharmacol. **49** (1996) 112.

Pimplikar, S.W. and K. Simons, J. Cell Sci. Suppl. **17** (1993) 27.

Popot, J. and D. Engelman, Biochemistry **29** (1990) 4031.

Rands, E., M.R. Candelore, A.H. Cheung, W.S. Hill, C.D. Strader, and R.A. Dixon, J. Biol. Chem. **265** (1990) 10759.

Rohrer, D.K. and B.K. Kobilka, Phys. Rev. **78** (1998) 35.

Samama, P., S. Cotecchia, T. Costa, and R.J. Lefkowitz, J. Biol. Chem. **268** (1993) 4625.

Schatz, J. and B. Dobberstein, Science **271** (1996) 1519.

Schwartz, T.W., U. Gether, H.T. Schambye, and S.A. Hjorth, Current Pharmaceutical Design **1** (1995) 355.

Shaw, G., Bioessays **18** (1996) 35.

Simon, M., M. Strathmann, and N. Gautam, Science **252** (1991) 802.

Sunahara, R.K., C.W. Dessauer and A.G. Gilman, Annu. Rev. Pharmacol. Toxicol. **36** (1996) 461.

Tsunoda, S., J. Sierralta, Y. Sun, R. Bodner, E., Suzuki, A. Becker, M. Socolich, and C.S. Zuker, Nature **388** (1997) 243.

van Biesen, T., B.E. Hawes, D.K. Luttrell, K.M. Krueger, K. Touhara, E. Porfiri, M. Sakaue, L.M. Luttrell, and R.J. Lefkowitz, Nature **376** (1995) 781.

Vanhoutte, P.M., P.P.A. Humphrey, and M. Spedding, Pharmacol. Rev. **48** (1996) 1.

Valdenaire, O. and P. Vernier, Prog. Drug Res. **49** (1997) 173.

Vernier, P., H. Philippe, P. Samama, and J. Mallet, in: Y. Pichon (ed.), *Comparative Molecular Neurobiology* (Birkauser, Zurich, 1993) pp. 297.

Vernier, P., B. Cardinaud, O. Valdenaire, H. Philippe, and J.D. Vincent, Trends Pharmacol. Sci. **16** (1995) 375.

von Heijne, G. and C. Manoil, Protein Eng. **4** (1990) 109.

Zuckerkandl, E., J. Mol. Evol. **39** (1994) 661.

SECTION IV. Lectures and Seminars Presented at the Summer School but not Published in the Proceedings

COURSE SUMMARY 1

THE ENDOMEMBRANE SYSTEM

David D. Sabatini

*NYU School of Medicine, Department of Cell Biology, 550 First Avenue,
New York, NY 10016, USA*

G. Zaccai, J. Massoulié and F. David, eds.
Les Houches, Session LXV, 1996
De la Cellule au Cerveau
From Cell to Brain: Intra- and Inter-Cellular Communication –
The Central Nervous System
© *1998 Elsevier Science B.V. All rights reserved*

195

Contents

1. Annotated bibliography of the course

A comprehensive review is to be found in the chapter by Sabatini and Adesnik in the 7th edition of the Metabolic and Molecular Bases of Inherited Disease. The appropriate reference is:

David D. Sabatini and Milton B. Adenisk, Biogenesis of membranes and organelles, in: C.R. Scriver, A.L. Beaudet, W.S. Sly, and D. Valle, eds, *The Metabolic and Molecular Bases of Inherited Disease*, 7th Edition, Vol. I (Mc-Graw-Hill, New York, 1995) pp. 459–553.

The following collection of articles, which appeared in Science in 1996 under the heading of Protein Kinesis, should be more than suficient to provide an advanced introduction to this topic. Those that relate directly to the subjects discussed in the course are marked by (*):

Dirk Görlich and Iain W. Mattaj, Nucleocytoplasmic transport, Science 271 (1996) 1513–1518.

(*) Gottfried Schatz and Bernhard Dobberstein, Common principles of protein translocation across membranes, Science 271 (1996) 1519–1526.

(*) Randy Schekman and Lelio Orci, Coat proteins and vesicle bulding, Science 271 (1996) 1526–1533.

(*) Pietro De Camilli, Scott D. Emr, Peter S. McPherson, and Peter Novick, Phosphoinositides as regulators in membrane traffic, Science 271 (1996) 1533–1539.

Richard B. Vallee and Michael P. Sheetz, Targeting of motor proteins, Science 271 (1996) 1539–1544.

An additional good review in another issue of Science is: James E. Rothman and Felix T. Wieland, Protein sorting by transport vesicles, Science 272 (1996) 227–234.

COURSE SUMMARY 2

THE HISTORY OF THE TWO-STAGE MODEL FOR MEMBRANE PROTEIN FOLDING

Donald M. Engelman

*Department of Molecular Biophysics and Biochemistry, Yale University,
266 Whitney Ave., CT 06520, New Haven, USA*

G. Zaccai, J. Massoulié and F. David, eds.
Les Houches, Session LXV, 1996
De la Cellule au Cerveau
From Cell to Brain: Intra- and Inter-Cellular Communication –
The Central Nervous System

201

Contents

1. Annotated bibliography relating to the course

The following is an annotated bibliography that charts the evolution of ideas, both correct and incorrect, that have led to the current state of the two-stage model for membrane protein folding. It is hoped that this exercise in the evolution of scientific ideas might be of interest to some of those who attended the course in Les Houches, and perhaps to others as well.

D.M. Engelman, R. Henderson, A.S. McLaughlin, and B.A. Wallace, Path of the polypeptide in bacteriorhodopsin, Proc. Natl. Acad. Sci. USA 77 (1980) 2023–2027.

When the sequence of bacteriorhodopsin became available, several at the MRC in Cambridge became interested in fitting it into the low resolution structural map obtained from electron microscopy. We peered at the sequence and developed a set of rough criteria, including the polarity of stretches that we thought might be embedded in the membrane as helices, the lengths of loops between these helices, the presumed interactions of polar groups to make salt bridges, and the projected densities of the helices. We developed a rank order of models, and the number one model has, in fact, proved to be correct. However, our assumptions about ion pairing turn out to be incorrect (with hindsight), so it was quite fortuitous that we nonetheless got the correct answer. The line of thinking about peptides in bilayers did, however, set the stage for further thinking.

D.M. Engelman and G. Zaccai, Bacteriorhodopsin is an inside-out protein, Proc. Natl. Acad. Sci. USA 77 (1980) 5894–5898.

Using neutron scattering and selective labeling of bacteriorhodopsin, we got a rough idea of the orientation of helices with respect to the bilayer facing and protein facing sides of the helical bundle in bacteriorhodopsin. This led to further thinking about the interactions with lipid and with other helices, mainly focusing on the idea that the lipid faces should be relatively non-polar.

D.M. Engelman and T.A. Steitz, The spontaneous insertion of proteins into and across membranes: The helical hairpin hypothesis, Cell 23 (1981) 411–422.

In this paper, a number of ideas were refined concerning the interactions of segments of polypeptides with lipid bilayers. That the fundamental premise (that the transmembrane secretion and membrane insertion of proteins can proceed by spontaneous insertion into lipid bilayers) is incorrect in the majority of cases has obscured the logic of other points made in the paper. Among these are the idea that the strength of hydrogen bonding in the lipid bilayer will dictate helical conformation, that hydrophobicity favors direct insertion of helical structures, and that progressive insertion of a polypeptide may be stopped by a hydrophobic helix (later called a stop transfer sequence). At about the same time, we began developing a scale to assess the free energy of insertion of different groups.

T.A. Steitz, A. Goldman, and D.M. Engelman, Quantitative application to the helical hairpin hypothesis to membrane proteins, Biophys. J. 37 (1982) 817–831.

In this article, we introduced the idea of scanning amino acid sequences for hydrophobic regions that might be transmembrane helices.

D.M. Engelman and T.A. Steitz, On the folding and insertion of globular membrane proteins, in: D. Wetlaufer, ed., *The Protein Folding Problem* (AAAS Publications, 1983) pp. 87–113.

Thinking further about the interactions of helices with bilayers led to the notion that a thermodynamic scale might be developed. It is first presented in this article, together with more ideas about the scanning of polypeptide sequences to find inserted helices.

D.M. Engelman, T.A. Steitz, and A. Goldman, Identifying non-polar transbilayer helices in amino acid sequences of membrane proteins, Ann. Rev. Biophys. Biophys. Chem. 15 (1986) 321–353.

Further analysis led to points made in this review about the testing of programs to identify transmembrane helices using known structures, which at that time included the newly determined structure of the photosynthetic reaction center. The conclusion was that the programs work reasonably well in identifying helical regions, and that many scales of hydrophobicity will work.

J.-L. Popot, S-E. Gerchmann, and D.M. Engelman, Refolding of bacteriorhodopsin in lipid bilayers: A thermodynamically controlled two-stage process, J. Mol. Biol. 198 (1987) 655–676.

This article, in fact, represents the direct experimental test of the two-stage model, which contends that helical structures can be regarded as independent, stable domains that interact side-to-side to form higher order structure in helical membrane proteins. The first stage was represented by two fragments of bacteriorhodopsin, one with two helices and one with five, separately incorporated into lipid vesicles, which were subsequently fused to allow the fragments to interact. The resulting structure bound retinal, and could reform the crystal lattice of purple membranes. While single helices were not tested, the main ideas of the two-stage folding notion are in this publication.

J.-L. Popot and D.M. Engelman, Membrane protein folding and oligomerization: The two-stage model, Biochemistry 29 (1990) 4031–4037.

In this article, the two-stage model is developed in a more formal way. In fact, the ideas that are presented here were mainly in place previously, but this is the most cited reference for the idea, and it is detailed fairly well. The main ideas are that individual helices are stabilized by the hydrophobic effect and hydrogen bonding in forming stable transbilayer structures, and that other interactions between the helices and lipid, between lipid and lipid and between the helices dictate the formation of higher order structure. At that time, we thought that the interaction of helices would be governed by the superior packing of helices with each other compared with their packing with lipid. While this may be an important energy term, the more important idea of the coordination of van der Waals interactions was not considered at this time.

T.W. Kahn and D.M. Engelman, Bacteriorhodopsin can be refolded from two independently stable transmembrane helices and the complementary five-helix fragment, Biochemistry 31 (1992) 6144–6151.

We finally succeeded in a more formal test of the idea that single helices could form independent structures that would interact with each other in a folding process. Here we show that the first helix, synthesized and inserted into lipid bilayer vesicles, added to a second population of vesicles containing the second helix as a separate entity, and in turn added to a third population of vesicles containing the remaining five helices will produce bacteriorhodopsin when the three vesicle populations are fused with each other and retinal is added. The individual vesicles were studied to show that they contain predominately α-helices.

M.A. Lemmon, J.M. Flanagan, J. Hunt, B.D. Adair, B.J. Bormann, C.E. Dempsey, and D.M. Engelman, Glycophorin A dimerization is driven by specific interactions between a-helices, J. Biol. Chem. 267 (1992) 7683–7689.

M.A. Lemmon, J.M. Flanagan, H.R. Treutlein, J. Zhang, and D.M. Engelman, Sequence specificity in the dimerization of transmembrane a-helices, Biochemistry 31 (1992) 12719–12725.

Using a chimeric construct of the transmembrane region of glycophorin A as a carboxyterminal extension of staphylococcal nuclease, we succeeded in studying the dimerization of glycophorin A helices in SDS micelles. An extensive mutagenesis study showed that the surfaces interact in a detailed fashion that is sensitive to small changes in amino acid structure at the interface.

P.D. Adams, D.M. Engelman, and A.T. Brunger, An improved prediction for the structure of the dimeric transmembrane domain of glycophorin A obtained through global searching, Proteins 26 (1996) 257–261.

Using simulated annealing and molecular dynamics, a search was conducted for favorable interactions between the glycophorin helices. A small number of favorable structures were found, one of which agreed with the mutagenesis study.

K.R. MacKenzie, J.H. Prestegard, and D.M. Engelman, A transmembrane helix dimer: Structure and implications, Science 16 (1997) 131–133.

The structure of the glycophorin helix dimer in detergent micelles was determined by NMR. The structure agreed quite well with the predicted model, but was substantially more detailed and differed in small ways. What it revealed is the very precise fit of two van der Waals surfaces, which appear to be minimally disturbed from those expected for the separate helices. Thus, the idea emerged that the energetics of the second stage of the two-stage model would involve the precise van der Waals fit of surfaces, and raised the notion that the formation of the surfaces in the first stage of the model contributes to the interaction energy by arranging an array of relatively fixed points in space for Van der Waals interactions to occur. Thus, the energy arises in part from the extended, pre-formed character of the surface with little contribution from entropy.

K.G. Fleming, A.L. Ackerman, and D.M. Engelman, The effect of point mutations on the free energy of transmembrane a-helix dimerization, J. Mol. Biol. 272 (2) (1997) 266–275.

Using analytical ultracentrifugation, we have developed a method for measuring free energies of association of helices in detergent environments. This method will, we anticipate, allow us to include an energy scale in our deliberations about helix interaction in the second stage of the two-stage model.

2. Conclusion

The evolution of concepts of helix stability in bilayers and helix–helix interaction embodied in the two-stage model has provided a useful pathway toward understanding membrane protein folding and stability. In the future, it is hoped that additional energy terms will be examined and understood, so that the more complex cases of polytopic proteins containing strongly polar groups in their transmembrane regions can be successfully analyzed. It is also hoped that functional states of membrane proteins can be understood in terms of alternative helix–helix interactions.

COURSE SUMMARY 3

POST SYNAPTIC RECEPTORS AND THE
ORGANISATION OF THE SYNAPSE

Regis Kelly

Hormone Research Institute, University of California,
San Francisco, CA 94743 05434, USA

G. Zaccai, J. Massoulié and F. David, eds.
Les Houches, Session LXV, 1996
De la Cellule au Cerveau
From Cell to Brain: Intra- and Inter-Cellular Communication –
The Central Nervous System

211

Contents

1. Bibliography relating to the course

1.1. Quantal release

H. von Gersdorff and G. Matthews, Dynamics of synaptic vesicle fusion and membrane retrieval in synaptic terminals. Nature 367 (1994) 735–739.

M. Frerking, S. Borges, and M. Wilson, Variation in CABA mini amplitude is the consequence of variation in transmitter concentration. Neuron 15 (1995) 885–895.

J.S. Isaacson and B. Walmsley, Counting quanta: direct measurements of transmitter release at a central synapse. Neuron 15 (1995) 875–884.

M. Lindau and W. Almers, Structure and function of fusion pores in exocytosis and ectoplasmic membrane fusion. Curr. Opinion Cell Biol. 7 (1995) 509–517.

1.2. The SNARE hypothesis

M.K. Bennett and R.H. Scheller, The molecular machinery for secretion is conserved from yeast to neurons. Proc. Natl. Acad. Sci. USA 90 (1993) 2559–2563.

M.K. Bennett, N. Calakos, and R.H. Scheller, Syntaxin-a synaptic protein implicated in docking of synaptic vesicles at presynaptic active zones. Science 257 (1992) 255–259.

M.K. Bennett, J.E. Carcia-Arraras, L.A. Elferink, K. Peterson, A.M. Fleming, C.D. Hazuka, and R.H. Scheller, The syntaxin family of vesicular transport receptors. Cell 74 (1993) 863–873.

N. Calakos, M.K. Bennett, K.E. Peterson, and R.H. Scheller, Protein–protein interactions contributing to the specificity of intracellular vesicular trafficking. Science 263 (1994) 1146–1149.

E. Ralston, S. Beushausen, and T. Ploug, Expression of the synaptic vesicle proteins VAMPs/synaptobrevins 1 and 2 in non-neural tissues. J. Biol. Chem. 269 (1994) 15403–15406.

A. DiAntonio, R.W. Burgess, A. Chin, D.L. Deitcher, R. Scheller, and T.L. Schwarz, Identification and characterization of Drosophila genes for synaptic vesicle proteins. J. Neurosci. 13 (1993) 4924–4935.

S. Ferro-Novick and R. Jahn, Vesicle fusion from yeast to man. Nature 370 (1994) 191–193.

T. Galli, T. Chilcote, O. Mundigl, T. Binz, H. Niemann, P. De Camilli, Tetanus toxin-mediated cleavage of cellubrevin impairs exocytosis of transferrin receptor-containing vesicles in CHO cells. J. Cell Biol. 125 (1994) 1015–1024.

Y. Hata, C.A. Slaughter, and T.C. Sudhof, Synaptic vesicle fusion complex contains unc-18 homologue bound to syntaxin. Nature 366 (1993) 347–351.

J. Pevsner, S.C. Hsu, and R.H. Scheller, n-/Sec1: a neural-specific syntaxin-binding protein. Proc. Natl. Acad. Sci. USA 91 (1994) 1445–1449.

J. Pevsner, S.C. Hsu, J.E.A. Braun, N. Calakos, A.E. Ting, M.K. Bennett, and R.H. Scheller, Specificity and regulation of a synaptic vesicle docking complex. Neuron 13 (1994) 353–361.

G. Schiavo, F. Benfenati, B. Poulain, O. Rossetto, P. Polverino de Laureto, B.R. DasGupta, and C. Montecucco, Tetanus and botulinum-B neurotoxins block neurotransmitter release by proteolytic cleavage of synaptobrevin. Nature 359 (1992) 832–835.

T. Sollner, M.K. Bennett, S.W Whiteheart, R.H. Scheller, and J.E. Rothman, A protein assembly–disassembly pathway in vitro that may correspond to sequential steps of synaptic vesicle docking, activation and fusion. Cell 75 (1993) 409–418.

1.3. Calcium regulation

A. DiAntonio, K.D. Parfitt, and T.L. Schwarz, Synaptic transmission persists in synaptotagmin mutants of Drosophila. Cell 73 (1993) 1281–1290.

J. Littleton, M. Stern, K. Schulze, M. Perin, and H. Bellen, Mutational analysis of Drosophila synaptotagmin demonstrates its essential role in Ca^{++} activated neurotransmitter release. Cell 74 (1993) 1125–1134.

M. Nonet, K. Ctundahl, B.J. Meyer, and J.B. Rand, Synaptic function is impaired but not eliminated in C. elegans mutants lacking synaptotagmin. Cell 73 (1993) 1291–1305.

A.C. Petrenko, M. Perin, B.A. Davletov, Y. Ushkaryov, M. Ceppert, and T.C. Sudhof, Binding of synaptotagmin to the α-latrotoxin receptor implicates both in synaptic vesicle exocytosis. Nature 353 (1991) 65–68.

1.4. Synaptic regulation

N. Brose, A.G. Petrenko, T.C. Sudhof, and R. Jahn, Synaptotagmin: a calcium sensor on the synaptic vesicle surface. Science 256 (1992) 1021–1025.

G. Fischer von Mollard, B. Stahl, C. Li, T.C. Sudhof, and R. Jahn, Rab proteins in regulated exocytosis. TIBS 19 (1994) 164–168.

P. De Camilli, S.D. Emr, P.S. McPherson, and P. Novick, Phosphoinositides as regulators in membrane traffic. Science 271 (1996) 1533–1539.

P.S. McPherson, E.P. Carcia, V.I. Slepnev, C. David, X. Zhang, D. Grabs, W.S. Sossin, R. Bauerfeind, Y. Nemoto, and P. De Camilli, A presynaptic inositol-65-phosphatase. Nature 379 (1996) 353–357.

C.F. Stevens and Y. Wang, Facilitation and depression at single central synapses. Neuron 14 (1995) 795–802.

Y. Coda and C.F. Stevens, Long-term depression properties in a simple system. Neuron 16 (1996) 103–111.

1.5. Synaptic vesicle biogenesis

P.J. Robinson, J.M. Sontag, J.P. Liu, E. Fykse, C. Slaughter, H. McMahon, and T.C. Sudhof, Dynamin GTPase regulated by protein kinase C phosphorylation in nerve terminals. Nature 365 (1993) 163–166.

W.J. Betz and L.-G. Wu, Kinetics of synaptic-vesicle recycling. Curr. Biol. 5 (1995) 1098–1101.

R.H. Scheller, Membrane trafficking in the presynaptic nerve terminal. Neuron 14 (1995) 893–897.

T.A. Ryan, S.J. Smith, and H. Reuter, The timing of synaptic vesicle endocytosis. PNAS 93 (1996) 5567–5571.

J.Z. Zhang, B.A. Davletov, T.C. Sudhof, and R.G.W. Anderson, Synaptotagmin I is a high affinity receptor for clathrin AP-2: implications for membrane recycling. Cell 78 (1994) 751–760.

P. De Camilli and K. Takei, Molecular mechanisms in synaptic vesicle endocytosis and recycling. Neuron 16 (1996) 481–486.

P. De Camilli, K. Takei, and P.S. McPherson, The function of dynamin in endocytosis. Curr. Opinion Neurobiol. 5 (1995) 559–565.

K. Takei, O. Mundigl, L. Daniell, and P. De Camilli, The synaptic vesicle cycle: a single vesicle budding step involving clathrin and dynamin. J. Cell Biol. 133 (1996) 1237–1250.

1.6. Secretory granule biogenesis

M. Crimes and R.B. Kelly, Intermediates in the constitutive and regulated secretory pathways released in vitro from semi-intact cells. J. Cell Biol. 117 (1992) 539–549.

A.S. Dittie, N. Hajibagheri, and S.A. Tooze, The AP-1 adaptor complex binds to immature secretory granules from PC12 cells and is regulated by ADP-ribosylation factor. J. Cell Biol. 132 (1996) 523–536.

S. Natori and W.B. Huttner, Chromogranin B (secretogranin I) promotes sorting to the regulated secretory pathway of processing intermediates derived from a peptide hormone precursor. PNAS 93 (1996) 4431–4436.

C. Wiedemann, T. Schafer, and M.M. Burger, Chromaffin granule-associated phosphatidylinositol 4-kinase activity is required for stimulated secretion. EMBO J. 15 (1996) 2094–2101.

Y.-G. Chen and D. Shields, ADP-ribosylation factor-1 stimulates formation of nascent secretory vesicles from the trans-Golgi network of endocrine cells. J. Biol. Chem. 271 (1996) 5297–5300.

1.7. Synapse adhesion

J.T. Campanelli, S.L. Roberds, K.P. Campbell, and R.H. Scheller, A role for dystrophin-associated glycoproteins and utrophin in agrin-induced AChR clustering. Cell 77 (1994) 663–674.

S.H. Gee, F. Montanaro, M.H. Lindenbaum, and S. Carbonetto, Dystroglycan-α, a dystrophin-associated glycoprotein, is a functional agrin receptor. Cell 77 (1994) 675–686.

Y.A. Ushkaryov, A.G. Petrenko, M. Ceppert, and T.C. Sudhof, Neurexins: synaptic cell surface proteins related to the α-latrotoxin receptor and laminin. Science 257 (1992) 50–56.

M. Kennedy, Biochemistry of synaptic regulation in the central nervous system. Ann. Rev. Biochem. 63 (1994) 571–600.

M.B. Kennedy, The postsynaptic density. Curr. Opinion Neurobiol. 3 (1993) 732–737.

M. Gautam, P.C. Noakes, J. Mudd, M. Nichol, G.C. Chu, J.R. Sanes, and J. Merlie, Failure of postsynaptic specialization to develop at neuromuscular junctions of rapsyn-deficient mice. Nature 377 (1995) 232–236.

H.-C. Kornau, L.T. Schenker, M.B. Kennedy, and P.H. Seeburg, Domain interaction between NMDA receptor subunits and the postsynaptic density protein PSD-95. Science 269 (1995) 1737–1740.

R.J. Kleiman and L.F. Reichardt, Testing the agrin hypothesis. Cell 85 (1996) 461–464.

STUDENT SEMINAR 1

INHIBITION AS BINDING CONTROLLER AT THE LEVEL OF A SINGLE NEURON (INFORMATION PROCESSING IN A PYRAMIDAL-TYPE NEURON)

Alexander K. Vidybida

Bogolyubov Institute for Theoretical Physics, Metrologichna str. 14-B, 252143, Kiev, Ukraine

G. Zaccai, J. Massoulié and F. David, eds.
Les Houches, Session LXV, 1996
De la Cellule au Cerveau
From Cell to Brain: Intra- and Inter-Cellular Communication –
The Central Nervous System

Contents

1. Introduction

The voltage-threshold principle [1] is widely accepted as criterion for a neuron triggering under external stimuli. Its adequacy is proven for step-like stimuli experimentally and theoretically [2]. On the other hand, a natural stimulus for a cortical neuron comprises up to several thousands EPSPs [3], and has a gradual time course. The applicability of the voltage threshold principle for this kind of stimuli is under question [4–6]. At the same time, it is suggested that exact timing of spikes in a neuronal assembly is essential for binding, or feature linking during processing of visual [7–9], auditory [10], or complex [11] sensory information. A definite timing of spikes results in a definite relative timing of EPSPs in a secondary neuron. Our purpose is to establish how does the ability to cause a spike depend on a relative timing of EPSPs within a compound stimulus.

2. Methods

The compound stimulus is taken in the form

$$\text{CompEPSP}(t) = \sum_{i=1}^{1000} \text{EPSP}(t - t_i), \tag{1}$$

where $\text{EPSP}(t)$ is the unitary EPSP time course; t_n, $1 \leq n \leq 1000$ are the random numbers within a fixed time window. We use the H-H set of equations [2] combined with the Monte Carlo algorithm to calculate numerically the dependence of firing probability on the window width, W, if the stimulus (1) is applied to a neuron [6]. A constant potassium conductance is added to the first H-H equation to account for the role of inhibition.

3. Results

Let us denote the degree of temporal coherence between the unitary inputs in the stimulus (1) by the inverse window width: $TC = 1/W$. The obtained dependences on TC of the firing probability are shown in Fig. 1. The four curves are calculated for inhibition potentials 0.0, 1.3, 3.8, 7.6 mV, respectively, from the left to the right. The step-like pattern suggests to a threshold-type principle.

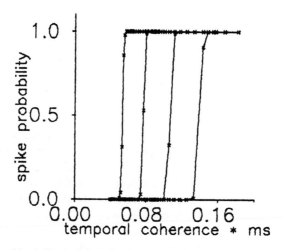

Fig. 1. Obtained dependencies on *TC* of the firing probability.

4. Conclusions and discussion

The results obtained lead to the following conclusions:

(1) A compound stimulus will trigger a neuron if and only if the degree of temporal coherence between the EPSPs within it is above a definite threshold. In other words, a neuron in natural conditions operates as time coherence discriminator.

(2) The degree of temporal coherence necessary for firing can be properly adjusted by means of inhibition.

The above conclusions allow one to offer the following account of the information processing in a single neuron: Synaptic inputs are considered as the inputs from some sensors, an initial segment of axon as the output to one or more effectors. The neuron receives signals into its inputs as signs of various events in the external region. A signal to the effectors has to be sent only if a definite compound event occurs. A decision that the required event has happened is made in the soma, the axonal hillock region, based on the degree of temporal coherence between the inputs.

In other words, the neuron converts a suitable set of elementary events represented by the set of synaptic inputs into a single event represented by the action potential in the axon. In this sense we say that binding appears as early as at the single neuron level, and inhibition effectively controls the conditions necessary for this kind of binding.

In this context, the quick process of synaptic transmission, which is characterized by the postsynaptic current (Fig. 2), may be treated as a signal from sensor.

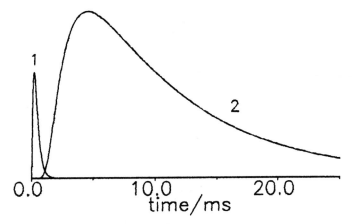

Fig. 2. Synaptic current (1) and excitatory postsynaptic potential (2) time courses. The current maximal value is equal to 400 nA, the potential maximal value in equal to 0.058 mV.

The slow transient process, characterized by the uEPSP, may be treated as an elementary short term memory mechanism. It is the slowness that makes the binding possible, because due to it the signals about present and recent past events are simultaneously existing in the neuron.

References

[1] D. Noble and R.B. Stein, J. Physiol. **187** (1966) 129.
[2] A.L. Hodgkin and A.F. Huxley, J. Physiol. **117** (1952) 500.
[3] P. Andersen, in: *Fidia Research Foundation Neuroscience Award Lectures* (Raven Press, New York, 1991) pp. 51.
[4] P.A. Kirkwood and T.A. Sears, J. Physiol. **322** (1982) 315.
[5] E.E. Fetz and B. Gustafsson, J. Physiol. **341** (1983) 387.
[6] A.K. Vidybida, Biol. Cybern. **74** (6) (1996) 539.
[7] R. Eckhorn, Biol. Cybern. **60** (1988) 121.
[8] C.M. Gray and W. Singer, Proc. Natl. Acad. Sci. USA **86** (1989) 1698.
[9] A.M. Sillito, Nature **369** (1994) 479.
[10] M. Joliot, U. Ribary, and R. Llinás, Proc. Natl. Acad. Sci. USA **91** (1994) 11748.
[11] A.R. Damasio, Neural Computation, **1** (1998) 123.

STUDENT SEMINAR 2

ISOLATED NERVE CELL RESPONSE TO LASER
IRRADIATION AND PHOTODYNAMIC EFFECT

A.B. Uzdensky

*Department of Biophysics and Biocybernetics, Physical Faculty, Rostov University,
Stachky avenue, 194/1 Rostov-on-Don, 344090, Russia*

G. Zaccai, J. Massoulié and F. David, eds.
Les Houches, Session LXV, 1996
De la Cellule au Cerveau
From Cell to Brain: Intra- and Inter-Cellular Communication –
The Central Nervous System

227

Contents

1. Introduction

Wide application of lasers in medicine vastly stimulated studies of visible-light effects on nonpigmented animal cells which do not contain specific light-absorbing molecules such as rhodopsin in retinal photoreceptor cells. Experimental data clarifying mechanisms of red laser radiation influence on living cells are presented by T.I. Karu in excellent reviews [1,2]. Electrophysiological and some cytological effects of visible laser radiation (mainly in the blue spectral region) on a single nerve cell and possible application of these data in testing of PDT photosensitizers will be considered in this paper.

Our main problems were:

(1) Photobiological problem – study of the mechanisms of laser radiation effects on nerve cells which are presumably common to various nonpigmented animal cells.

(2) Neurophysiological problem – elucidation of relations between subcellular structures selectively altered by laser radiation and nerve cell function.

(3) Application problem – testing and investigation of new PDT photosensitizers.

2. Object and methods

A suitable object for the study of the outlined problems is a slowly adapting crayfish stretch receptor neuron: a large single neuron capable to long-lasting firing with a nearly constant rate. At this stable background one can precisely investigate initial shifts of neuronal membrane state, dynamics of bioelectric changes, and terminal events leading to cell death. It is a classic neurophysiological object, and its structure, electrophysiological and biochemical features are well studied [3,4].

Neuron spikes were derived extracellularly by glass pipette electrodes filled with a saline and attached by vacuum to the axon. The spikes were amplified and their frequency was measured by an analog frequency meter and continuously recorded by a chart-recorder. At the beginning of the experiments we did set the initial firing frequency level at about 10–14 Hz by an appropriate receptor muscle extension. After stable control spike generation with this frequency during 15–30 min, various neuron functionally specialized regions were irradiated with a 8 μm laser microbeam focused by a microscope. The laser radiation sources used were helium–cadmium (441.6 nm, 1 kW/cm^2) and helium–neon (632,8 nm) lasers. Action spectra were studied using a dye laser pumped by a nitrogen laser (434–600 nm; 15 ns; 400 Hz, mean power 1 mW, mean intensity 1 kW/cm^2).

Neuron responses to laser microirradiation were modified by bioenergetic inhibitors: 2,4-dinitrophenol (0.1 mM), sodium azide (2–5 mM), sodium monoiodacetate (0.5–1 mM); photosensitizers: sodium fluoresceine (1 μg/ml) or janus green B (7 μg/ml); different calcium ion concentrations in saline (1/3- and 3-fold). In photosensitizers comparison experiments we stained neurons 30 min with methylene blue (50–500 nM), chlorin e6 (50–500 nM), or janus green B (1–10 μM), and then studied their responses to helium–neon laser irradiation. The used modificator concentrations did not significantly change the neuron firing frequency themselves. After control exposure of the neuron in modified saline and after firing frequency registration cells were irradiated. The parameters of the neuron response dynamics were measured and compared with unmodified ones by use of the standard Student's t-test or nonparametric Wilkokson test [5].

3. Single neuron response to blue laser microirradiation

The firing of a single nonpigmented nerve cell was insensitive to red laser radiation [6], but changed under blue laser irradiation [7,8]. In most of the experiments neurons polyphasically responded to long-lasting blue laser microirradiation: after some latency (L) the firing frequency increased (phase I), then gradually decreased (phase II), increased again (phase III), and reaching a relatively high level (30–50 Hz) the neuron firing abruptly and irreversibly ceased (phase IY). For quantitative analysis we characterized every response phase with the set of parameters: durations, t_i, values, Δf_i, and rates, v_i, of impulse frequency changes, where the subscript 'i' is the phase number (Fig. 1) [8].

Under some experimental conditions this EIE-type (excitatory–inhibitory––excitatory) response dynamics was modified: its different phases might be more prominent, or reduced, or absent. We compared modified and unmodified responses. We observed sometimes excitatory response (E-type) with a spike cessation occurring immediately after the phase I. Such E-type neuron responses were often recorded in winter, whereas full EIE-responses were usually observed in summer and autumn.

Taking into account the assumption on the inhibitory role of calcium ions in neuron response to laser irradiation (it is discussed below), these response changes may be related to season changes of calcium concentration in crayfish organism.

Sometimes the neuron response consisted of excitatory and next inhibitory

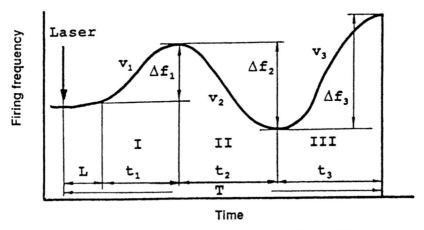

Fig. 1. Schematic representation of neuron response to laser microirradiation and denotements used. Ordinate: impulse frequency, abscissa: time.

phases (EI-type). This was observed when the inhibition process in phase II was so strong that the neuron was silent for a long time. This inhibition phase was presumably due to neuron hyperpolarization because this cell retained spike generation ability and fired under additional receptor muscle extension which generated an additional depolarizing receptor potential [7].

It is logical to assume that complex EIE-type dynamics is the superposition of two oppositely directed processes: depolarization with neuron firing acceleration, and hyperpolarization developing some time later. Inhibitory resources causing phase II are thought to be limited, and after their exhaustion the uninterruptedly acting excitatory factors may dominate and accelerate neuron firing again. To prove this scheme, we modified neuron response to laser microirradiation with different chemicals and investigated response changes. Experimental data are summarized in Table 1.

First indications of possible laser microirradiation targets in a neuron were obtained in the experiment with microirradiation of different functionally specialized cell regions [8,9]. It was shown that the cytoplasm area between nucleus and axon (C-zone), which is very rich in mitochondria, is most sensitive to laser microirradiation. Considering neuron response dependence on irradiation place, one could see two dependence types. In the first one some response phases (they were excitatory phases I and III) were reduced when the irradiation point moved in the axon-dendrite direction. Neuronal membrane excitability is also reduced in this direction, and we suspected that the laser effect on the neuronal membrane was the reason of these phases. In another response type the second inhibitory phase

Table 1

Modifications changing different phases of neuron response to laser microirradiation of its cytoplasm region

Modificator	Neuron response phases					
	Latency	Phase I	Phase II	Phase III	Lifetime	Reference
Season changes:						
summer	n.c.	n.c.	inc.	n.c.	inc.	15
winter	n.c.	n.c.	el.	el.	dec.	15
Different localization: [a]						
axon	inc.	n.c.	dec.	n.c.	dec.	8
nucleus	n.c.	n.c.	n.c.	dec.	n.c.	8
dendrite	inc	dec.	dec.	dec.	dec.	8
2,4-dinitrophenol	dec.	inc.	dec.	inc.	dec.	8
Sodium fluorescein	dec.	inc.	dec.	inc.	dec.	8
Janus green B	n.c.	n.c.	inc.	n.c.	inc.	8
Sodium monoiodacetate	dec.	inc.	el.	el.	dec.	8
Sodium azide	dec.	inc.	el.	el.	dec.	8
External Ca content:						
1/3-fold	n.c.	inc.	el.	el.	dec.	8
3-fold	n.c.	dec.	inc.	el.	inc	8
Dibunol	n.c.	dec.	n.d.	n.d.	inc.	9
D-mannit	n.c.	n.c.	n.c.	n.c.	n.c.	9
NaI	n.c.	n.c.	inc.	el.	inc.	9

n.c. = no significant changes, inc. = increase, dec. = decrease, n.d. = no data, el. = phase elimination.
[a] in comparison with C-zone irradiation.

was more prominent at neuron soma (C-zone, especially) irradiation as compared to dendrite or axon irradiation. The content of various intracellular organelles in neuron soma is much higher than in axon or dendrites. Because light absorbing molecules such as cytochromes and flavins are concentrated in mitochondria, we assumed that inhibition phase II was the result of a laser effect on mitochondria. We tried to prove these assumptions.

Specific photosensitization of different cellular structures with supravital stains significantly increases selectivity of laser action [10]. Photosensitization of neuronal membrane with fluorescein molecules which were adsorbed at cellular surface and could not penetrate into the cell due to negative charge (it was proved by fluorescent microscopy), caused an about 5-fold decrease of latency L and neuron lifetime T, and an 8- and 2-fold increase of frequency acceleration rates v_1 and v_3, respectively. However, inhibition phase II was reduced and inhibition rate v_2 was decreased by 1.5 times [9]. These data support our assumption that firing acceleration phases I and III were caused by laser radiation influence on the neuronal membrane.

Janus green B selectively staining mitochondria in living cells [11] vastly increased the photosensitivity of nonpigmented neurons. Even red helium–neon laser radiation, which could not change nerve cell firing itself, very effectively influenced firing of a stained cell [6]. Focusing the laser microbeam into the nucleus of a janus green stained neuron we observed E-type responses in most of the experiments, whereas EIE responses were observed when the C-zone was irradiated [9]. Considering that the vertical section of the laser microbeam focused into different cellular sites, one can see that the light flux through the neuronal membrane is almost the same in both cases. However, the number of irradiated mitochondria at microbeam focusing on the nucleus center is markedly less than in the case of focusing on cytoplasm because only thin cytoplasm layers containing mitochondria above and under the nucleus were irradiated in this case, and the total "mitochondrial effect" has to be much less.

A similar difference between blue laser microirradiation of C-zone and nucleus was observed in the case of response modification by 2,4-dinitrophenol. It is a well known uncoupler of oxidative phosphorylation in mitochondria. But in our experiments the photodynamic effect was thought to be more important: this yellow substance well absorbing blue laser light intensified all response stages. However, at irradiation of the nucleus, phase II was much less than in the case of C-zone irradiation [8,9]. These findings prove that laser influence on mitochondria accounts for inhibition phase II in neuron response.

Using inhibitors selectively acting on different components of the bioenergetic chain one can change their redox-states and hence optical properties. Both, sodium azide inhibiting cytochrome C oxidase, the terminal point of respiration chain, and glycolytic inhibitor sodium monoiodacetate preventing electron supply to initial points of this chain produced polyphasic changes of neuron impulse activity themselves at concentrations more than 10 and 4 mM, respectively. The dynamics of these changes resembled laser-induced neuron response: rise of firing frequency (without latency), then decrease, repeated increase and abrupt cessation of impulse generation. It seems that points of action of different bioenergetic inhibitors and blue laser radiation are the same, and bioenergetic processes are involved in neuron response to laser microirradiation. At smaller concentrations sodium azide (2–5 mM) or sodium monoiodacetate (0.5–1 mM) did not change neuron firing itself but modified neuron response to laser microirradiation. Both inhibitors reduced or eliminated inhibition phase II in the most of the experiments [8,9]. The similarity of effects observed in both cases differing in their influence on redox states of the electron transfer chain components indicated that this difference and consequently the difference in optical absorption was not very significant and inhibition of energy production was more important for phase II emergence.

Electron-microscopic experiments showed that blue laser microirradiation of neuron cytoplasm affected mitochondria more significantly than other organelles. We observed their swelling, disruption of cristae and partial or complete loss of matrix. Inner mitochondrial membranes containing main redox systems were much more damaged than outer membranes. The degree of mitochondrial lesion depended on the phase of neuron response to laser irradiation [8,12].

What molecular components did serve as blue light photoreceptors in a non-pigmented nerve cell? Cytochemical study revealed drastic inactivation of the mitochondrial flavoprotein enzyme succinate dehydrogenase in the irradiated region of the neuron [14]. At the same time, cytochrome C oxidase, electron-transfer-ring hemoprotein integrated into inner mitochondrial membrane, was insensitive to blue laser radiation [15].

Action spectra of neuron response to C-zone microirradiation in the wavelength range from 434 to 600 nm revealed a rather sharp maximum at 460 nm for phase I parameters: L, t_1, and v_1. Absorption spectra with these maxima are characteristic for flavin compounds, and we suspected them as primary photoreceptors in a neuron. Action spectra of phase II parameters, t_2 and v_2, were not so sharp and contained an additional maximum at 488 nm implying additional component involvement in phase II emergence [8,13].

From these cytological data showing laser microirradiation induced mitochondrial lesion we could not understand how the blue laser light absorbed by mitochondrial flavins might evoke firing acceleration in phase I of cell response, which we ascribed to neuronal membrane injury. However, cytochrome-flavin redox chains are found not only in mitochondrial but also in plasma membrane and other intracellular membranes [16]. According to W. Schmidt considering primary physiological blue light action [17]: "a uniform picture appears to crystallize concerning the primary reactions: (1) The photoreceptor pigment is a flavin, probably a flavoprotein, (2) the photoreceptor is bound to a membrane, often the plasma membrane, in a highly dichroic manner, and (3) the primary reaction is a redox reaction". It is likely that in our case flavins bound to neuronal membrane are the primary blue light photoreceptors that caused membrane lesions, depolarization, and spike firing acceleration under laser irradiation.

In an attempt to reveal the role of some primary photochemical processes in neuron response to laser microirradiation we modified it with different quenchers of intermediate photochemical products. It is known that flavin photooxidation produces superoxide anion and hydrogen peroxide [17] promoting free radical lipid peroxidation and destroying biomembranes [18]. In our experiments the antioxidant dibunol (0.02 mM) protected the nerve cell against blue laser radiation injury [19]. It slowed down frequency changes in acceleration phase I and increased neuron lifetime. This proves that firing acceleration was due to free radical lipid peroxidation. (This experiment was carried out in winter when phase II was eliminated probably due to seasonal shifts of animal state, and unfortunately we could not say anything about free radical peroxidation involvement in this inhibitory process.)

Hydroxil radicals are known as very active intermediate products in ionizing radiation cell injury [20]. In our attempt to reveal the possible role of hydroxyl radicals in neuron response to blue laser microirradiation using their quencher D-mannit (10 mM) [21] we did not observe significant changes of different response phases. It seems that hydroxyl radicals, very important in radiobiological tissue injury, were not significant in blue laser radiation effect on neuron firing.

Observation of flavin-like action spectrum forced us to modify the neuron response to laser microirradiation with iodide-anion – a known flavin triplet quencher [22]. Our attempt to protect the neuron using sodium iodide (10 mM) was not successful. Although neuron lifetime was elongated by 1.4 times, it was not the actual protection because it was the result of significant phase II increase. Inhibition process was so prominent that EI-responses were observed in one-half of the experiments. We have no unequivocal explanation of this fact. It may be possible that at the concentration used (10 mM) iodide ions have a broad field of activity and inhibited secondary processes rather than quenched flavin triplet states [22]. They might inhibit processes leading to firing acceleration under irradiation, and therefore phase III could not develop after the inhibition phase in some experiments.

One another problem: how mitochondrial damage evokes inhibition of cell membrane activity? The existence has to be assumed of some mediator releasing from damaged mitochondria, diffusing to the cellular membrane, and hyperpolarizing it. We suggested that Ca ions are most preferable for this role. It is well known that they are actively accumulated in mitochondria and released after mitochondrial lesion [23]. They may hyperpolarize neurons and inhibit their firing through increase of membrane potassium conductance and threshold of spike generation [24,25]. In our experiments 3-fold calcium concentration slightly decreased the initial firing frequency level, presumably due to neuron hyperpolarization. It also reduced frequency shifts in acceleration phase I of neuron response and vastly enhanced inhibition phase II inducing firing cessation for 10–20 min or more. In this case we observed EI-type response in which phase III was eliminated and additional receptor muscle extension was necessary to resume firing. On the contrary, 1/3-fold calcium concentration itself markedly enhanced neuron excitability and increased initial frequency level from 10 to 14 Hz. In these conditions the phase I of neuron response was nearly twice as intensive as at normal calcium concentration, inhibition phase II was eliminated, and we observed

E-type response in almost all experiments. Hence, calcium ions were necessary for an inhibition phase to emerge [8,9]. These data prove the suggestion that calcium ions evoke inhibition of neuron firing under laser irradiation. It seems that after mitochondrial injury by laser microbeam calcium ions released from lesioned mitochondria, diffused to neuronal membrane, and inhibited neuron firing. This suggestion is supported by our finding that blue laser radiation (488 mn) inactivates ATPase activity in brain homogenates, and especially Ca-ATPase involved in Ca accumulation in mitochondria [8].

Collecting these data we propose the following mechanism of blue laser radiation effect on a nerve cell. We suppose that this mechanism is photochemical but not thermal because of long-lasting latency of firing changes (minutes and tenth of minutes at low irradiation intensity) and the lack of coagulation material at electron-microscopic photographs [8,12]. It seems that laser radiation absorbed mainly by flavins bound to cellular membranes induces lipid peroxidation in membranes directly or through superoxide anion formation [17]. Due to free radical damage, ion permeability of neuronal membrane increases, and the subsequent depolarization evokes spike acceleration. Some later inner mitochondrial membranes become injured, and free calcium ions released from lesioned mitochondria diffuse to neuronal membrane and hyperpolarize it. This inhibition process overcomes the excitatory effect of direct neuronal membrane damage and induces firing slowing down. After exhaustion of the Ca^{2+} store in irradiated mitochondria, the depolarizing effect of uninterrupted neuronal membrane damage due to flavin-mediated free radical injury becomes dominant again, and the acceleration phase reappeares. Firing acceleration goes on untill enhanced depolarization destructs the spike-generating mechanism as in the case of the cathodic block process. We suppose that these reactions are common to different nonpigmented animal cells responding to laser light.

4. Neurophysiological conclusion

Selective action of light microbeam on different cellular structures with parallel investigation of structural, metabolic, and functional shifts may serve as a useful tool in studies of the role of these structures in cell functions. So, elucidation of mechanisms underlying the information processing in a nerve cell and producing output impulse activity pattern, i.e. its integrative function, is of major neurophysiological importance. It is not clear yet, how this process proceeds: due to simple summation of excitatory and inhibitory input potentials at the neuronal membrane, or with an active participation of some intracellular organelles correcting output signals [26]. A single crayfish neuron is a simple and suitable model for study of this problem because only one input signal remains after its isolation – the exci-

Table 2

Dependencies of single stretch receptor neuron lifetime T and rate of firing acceleration v under photodynamic effect on photosensitizer concentration C

Photosensitizer	Concentration range (μM)	Regression equations	
		Acceleration rate, v	Lifetime, T
Chlorin e6	0.05–0.5	$\log v = 13.9 + 2.08 \log C$	$\log T = -12.3 - 2.02 \log C$
Methylene blue	0.05–0.5	$\log v = 9.9 + 1.42 \log C$	$\log T = -11.8 - 1.86 \log C$
Janus green	1–10	$\log v = 2.7 + 0.45 \log C$	$\log T = -0.55 - 0.36 \log C$

tatory receptor potential induced by receptor muscle extension. Fixing this input and selectively influencing mitochondria with a laser microbeam we showed that mitochondrial damage may induce inhibition of neuron firing mediated by calcium ions released from injured mitochondria [9]. It is evident that mitochondria may also influence the neuron firing pattern in some other situations.

5. Possible application of stretch receptor neuron for PDT photosensitizers testing

We used stretch receptor neuron for testing and comparison of new photosensitizers synthesized for photodynamic therapy of tumors (PDT). Using this preparation one can get information on initial cell function alterations, continuously follow cell response dynamics, and reveal cytotoxic events leading to cell death.

Isolated neurons were stained with various photosensitizers: methylene blue, a well known photogenerator of singlet oxygen; the effective PDT photosensitizer chlorin e6, and janus green B, selectively staining mitochondria in living cells. Then they were irradiated by helium–neon laser (632.8 nm, 0.3 W/cm^2). Neuron electrophysiological responses were rather simple: firing acceleration and subsequent irreversible block of spike generation, i.e. E-responses. We denoted firing acceleration as v, and neuron lifetime as T. Dependencies of v and T on photosensitizer concentrations C were well approximated by a linear function in double logarithmic coordinates, i.e. concentration dependencies were:

$$v = kC^L; \quad T = nC^M.$$

Regression equation parameters k, n, L, and M were estimated by the least-squares method (Table 2).

It is interesting that exponents L and M for chlorin e6 and M for methylene blue were close to 2. This means that interaction of two dye particles is necessary for elementary act leading to firing changes and subsequent cell death. At the same time mechanism of janus green B photodynamic action is probably other because L and M exponents were less than 1: 0.45 and 0.36, respectively. These data showed that an isolated stretch receptor neuron is very sensitive to photodynamic injury. It responded to as low a dye concentration as 50 nM. Photodynamic efficiencies of chlorin e6 and methylene blue were almost the same, but janus green B was much less effective. We are sure that this non-traditional model object is very suitable and useful for examination, testing and comparison of new PDT photosensitizers.

References

[1] T.I. Karu, *Photobiology of Low-Power Laser Therapy* (Harwood Academic, London, 1989).

[2] T.I. Karu, in: S.L. Jacques and A. Katzir (eds.), *Laser-Tissue Interaction YI*, Proc SPIE 2391 (1995) pp. 576.

[3] O.B. I'linsky, *Physiology of Sensory Systems. Part III. Physiology of Mechanoreceptors* (Nauka, Leningrad, 1975) (in Russian).

[4] E.E. Giacobini, in: *Neurosciences Research. Part II. Invertebrate Nerve Cell* (Academic Press, New York, 1969) pp. 111.

[5] B.M. Vladimirsky, *Mathematical Methods in Biology* (Rostov University Press, Rostov-on-Don, 1983) (in Russian).

[6] A.B. Uzdensky, Biol. Sci. **5** (1975) 45 (in Russian).

[7] A.B. Uzdensky, Biol. Sci. **3** (1980) 20 (in Russian).

[8] A.B. Uzdensky, in: S.L. Jacques (ed.), *Laser- Tissue Interaction IY*, Proc. SPIE 1882 (1993) pp. 254.

[9] A.B. Uzdensky, in: Fercher et al. (eds.), *Microscopy, Holography, and Interferometry*, Proc SPIE 2083 (1994) pp. 225.

[10] A.B. Uzdensky, Cytology **4** (1982) 1119 (in Russian).

[11] A. Lazarow and S.J. Cooperstein, J. Histochem. Cytochem. **1** (1953) 234.

[12] G.M. Fedorenko and A.B. Uzdensky, Cytology **28** (1986) 512 (in Russian).

[13] E.A. Morozova, V.V. Savransky, A.B. Uzdensky, and V.N. Gusatinsky, in: *Automatization in Cytological Investigations* (Naukova Dumika, Kiev, 1985) pp. 98 (in Russian).

[14] A.B. Uzdensky, Cytology **29** (1987) 1392 (in Russian).

[15] A.B. Uzdensky, *Electrophysiological and cytological study of isolated neuroreceptor cell response to total and local laser irradiation*, Dissertation, (Rostov University, Rostov-on-Don, 1980) (in Russian).

[16] H. Low and F.L. Crane, Biochim. Biophys. Acta **515** (1978) 141.

[17] W. Schmidt, in: H. Senger (ed.), *Blue Light Effects in Biological Systems* (1984) pp. 81.

[18] E.B. Burlakova, A.V. Alekseenko, et al., *Bioantioxydants in Radiation Damage and Malignant Growth* (Nauka, Moscow, 1975) (in Russian).

[19] A.B. Uzdensky, in: G. Delacretaz et al. (eds.), *Laser Interaction with Hard and Soft Tissues, IL*, Proc. SPIE, 2323 (1994) pp. 491.

[20] N.V. Timofeev-Ressovsky, A.V. Savich, and M.I. Shalnov, *Introduction in Molecular Radiobiology* (Medicina, Moscow, 1981) (In Russian).

[21] L.D. Luk'yanova, B.S. Balmuhanov, and A.T. Ugolev, *Oxygen-dependent Processes in Cell and its Functional State* (Nauka, Moscow, 1982) (In Russian).

[22] E. Schonbohm and E. Schonbohm, in: H. Senger (ed.), *Blue Light Effects in Biological Systems* (Springer Verlag, Berlin, Heidelberg, 1984) pp. 137.

[23] T. Pozzan, R. Rizutto, P. Volpe, and J. Meldolesi, Physiol. Rew. **74** (1994) 595.

[24] R.W. Meech, Comp. Biochem. Physiol. **42A** (1972) 493.

[25] R.W. Meech, Comp. Biochem. Physiol. **48A** (1974) 387.

[26] P.K. Anokhin, Usp. Fiziolog. Nauk **5** (1974) 5 in Russian).

SHORT REPORTS ON PH.D. STUDENT SEMINARS

G. Zaccai, J. Massoulié and F. David, eds.
Les Houches, Session LXV, 1996
De la Cellule au Cerveau
From Cell to Brain: Intra- and Inter-Cellular Communication –
The Central Nervous System

Contents

A number of Ph. D. students attended the summer school and gave brief reports on their on-going research, which provided the basis for exciting discussions. We thought it interesting to contact these students in order to find out how the work has developed, more than one year later. The replies follow.

The Editors

1. Structure and hydration of bacteriorhodopsin in its M-state studied by neutron diffraction

Martin Weik

Laboratoire de Biophysique Moléculaire, Institut de Biologie Structurale,
41, Avenue des Martyrs, 38027 Grenoble Cedex 1, France

This work was published in 1998. It is part of my Ph.D. thesis, which I presented in April 1998.

M. Weik, G. Zaccai, N.A. Dencher, D. Oesterhelt, and T. Hauß, Structure and hydration of the M-state of the bacteriorhodopsin mutant D96N studied by neutron diffraction, J. Mol. Biol. 275 (1998) 625–634.

2. Membranes, vesicles and micelles – a density functional approach

Carlos Rascon Diaz

Dpto. Fisica Teorica de la Materia Condensada, Facultad de Ciencas, CV 28049 Madrid, Spain

The first microscopic approach to the formation of membranes, vesicles and micelles was presented. Amphiphilic molecules were modelized by a spherical hard core plus an anisotropic interaction. The model was solved through density functional theory in a perturbative way, and showed striking results.

A. Somoza, E. Chacon, L. Mederos, and P. Tarazona, A model for membranes, vesicles and micelles in amphiphilic systems. J. Phys. Condens. Matter, 7 (1995) 5753–5776.

3. Elastic properties of the *Listeria Moncytogenes* tail

Fabien Gerbal

Inst. Curie, Section de Recherche URA 1379, 11 Rue Pierre et Marie Curie, 75231 Paris Cedex 05, France

The bacterium *Listeria Monocytogenes* is a model for the understanding of cell motility based on actin polymerization. Its membrane protein, Acta, is responsible for a local shift of the equilibrium in the action of actin polymerisation. This leads to the assembly of filaments from globular actin present in the infected medium. Filaments are reticulated at one pole of the bacterium and organized in a tubular structure: this "tail" is up to several hundred microns long and about one micron in diameter, comparable to the size of the bacterium. Steady addition of actin filaments at the extremity of the bacterium leads to the continuous building of this tail, functioning as an anchor in the medium in order to propel the bacteria forward. Since the protein organization at the molecular level has not yet been fully understood, we privileged a mesoscopic approach to the mechanism of motion. At such a scale, elastic properties of the tail are of primal interest. To achieve such a measurement, an optical tweezer producing forces of few tens of pico-Newtons on latex microspheres was used. By sticking such spheres on the tail we were able to bend and probe the tail elasticity. This data should reveal information on the the tail structure and help to discriminate between several theoretical models of *Listeria* motion.

4. Introduction to indirect detected ^{13}C NMR imaging and spectroscopy

Michael Heidenreich

Sektion Kernresonanzspektroskopie, University of Ulm, 89069 Ulm, Germany

C. Kunze and R. Kimmich, Proton detected ^{13}C imaging using cyclic J cross polarization, Magn. Reson. Imaging 12 (1994) 805.

M. Heidenreich, W. Köckenberger, R. Kimmich, N. Chandrakumar, and R. Bowtell, Investigation of Carbohydrat Metabolism and Transport in Castor Bean Seedlings by Cyclic J Cross Polarization Imaging and Spectroscopy, J. Magn. Reson. (1997) submitted.

M. Heidenreich, A. Spyros, W. Köckenberger, N. Chandrakumar, R. Bowtell, and R. Kimmich, CYCLCROP mapping of ^{13}C labelled compounds: Perspectives in polymer science and plant physiology, submitted to: *Spacially Resolved Magnetic Resonance: Methods and Applications in Material Science, Agriculture and*

Biomedicine, Eds. B. Bluemich, P. Bluemler, R. Botto, and E. Fukushima (VCH, Weinheim).

5. Crystallographic studies of the small ribosomal 30S subunit from *Thermus thermophilus* and bovine pancreatic trypsin – contrast variation and phasing with anomalous dispersion of phosphorus and sulfur

Sigrid Stuhrmann

Present address: Department of Biochemistry, University of Cambridge, Old Addenbrooke's Site, 80 Tennis Court Road, Cambridge CB2 1GA, UK

The presented work exploits the anomalous scattering of sulfur and phosphorus at their K-absorption edges ($\lambda_{K(Phosphorus)}$: 5.78 Å, $\lambda_{K(Sulfur)}$: 5.02 Å) in crystallographic studies of biological macromolecules. The measurements were made with a specially designed instrument for soft X-ray diffraction. Due to increased absorption effects with soft X-rays the whole instrument is evacuated. For crystallographic studies of ribosomal crystals the following technical improvements have been achieved:

– The increased radiation damage of biological material in the soft X-ray range has been reduced by implementation of a cooling device consisting of a 5-stage Peltier cascade which cools to $-110°C$ making first quantitative MAD data collections possible.

– The crystal mounting and freezing technique of small and very sensitive ribosomal crystals has been adapted to the special environment. For test purposes crystals of bovine trypsin were used to develop the mounting and freezing technique.

– A vacuum tight sample cell with minimized X-ray absorption for ribosomal and protein single crystals has been developed.

Bovine trypsin (m.w.: 24 kDa) with its sulfur containing amino acids (2 methionines and 6 cystines), the structure of which is well known, was used as a test candidate. Data were collected near the K-absorption edge of sulfur. It was shown that anomalous difference Patterson maps calculated from the experimental data and the known model are correlated. An electron density map at 5 Å resolution was derived from the measured anomalous data with the aid of the coordinates of the sulfur atoms from the known model.

The diffraction patterns of trypsin show significant effects of anisotropy of anomalous scattering (AAS) due to the sulfur containing amino acids. For further analysis a plausible model spectrum of the AAS of disulfide bridges was developed.

The high amount of ammonium sulfate (2.5 M) in the mother liquor of bovine trypsin crystals offers the opportunity to apply the method of anomalous contrast variation near the K-edge of sulfate ions (oxidation state of sulfur: +6), which is shifted by 10 eV to higher energies with respect to the K-edge from the sulfur containing amino acids (oxidation state of sulfur: -2). Anomalous diffraction data yielded an anomalous contrast of 0.02 $e^-/Å^3$ in the real part, which amounts in an unusually high fraction of 50% of the non-resonant contrast. The measurement of trypsin at the sulfate K-edge served as useful prerequisite for the study of single crystals of ribosomal particles near the K-edge of phosphorus.

Anomalous diffraction data of the small ribosomal 30S subunit from *Thermus thermophilus* were collected near the K-absorption edge of phosphorus (λ_K: 5.78 Å) at three or five wavelengths respectively. These crystals diffract short wavelength X-rays to about 7 Å, and the soft X-ray radiation up to 30 Å Bragg resolution. This is significant for low resolution structural investigations based on the concepts of contrast variation, where the 1515 phosphorus atoms provide an adequate description of the rRNA phase. The average contribution of anomalous dispersion amounts to almost 48% of the non-resonant diffracted intensity. The real part of the anomalous contrast reaches a value of 0.06 $e^-/Å^3$ in the RNA phase which is comparable with neutron scattering experiments using isotopic substitution.

S. Stuhrmann, M. Hütsch, C. Trame, J. Thornas and H.B. Stuhrmann, Anomalous dispersion with edges in the soft X-ray region: First results of diffraction from single crystals of ribosomes near the K-absorption edge of phosphorus, J. Synchrotron Rad. 2 (1995) 83–86.

S. Stuhrmann, K.S. Bartels and H.B. Stuhrmann, Towards the structure determination of proteins from the near edge anomalous dispersion of sulphur: a comparison of the first results from trypsin with the known structure, Zeitschrift für Kristallographie, 212 (1997) 350–354.

S. Stuhrmann, K.S. Bartels, W. Braunwarth, R. Doose, F. Dauvergne, A. Gabriel, A. Knöchel, M. Marmotti, H.B. Stuhrmann C. Trame and M.S. Lehrnann, Anomalous dispersion with edges in the soft X-ray region: First results of diffraction from single crystals of trypsin near the K-absorption edge of sulfur, J. Synchrotron Rad. 4 (1997) 298–310.

SECTION V. CONCLUSION

RELATIONS BETWEEN PHYSICS AND BIOLOGY

Bernard Jacrot

Rue de la Place, 84160 Cucuron, France

G. Zaccai, J. Massoulié and F. David, eds.
Les Houches, Session LXV, 1996
De la Cellule au Cerveau
From Cell to Brain: Intra- and Inter-Cellular Communication –
The Central Nervous System

Contents

255

1. Relations between physics and biology

Physics and Biology both belong to the natural sciences and as such have developed through the use of common scientific methods. They are, however, at different stages of development. Physics really started as a science with Galileo and Newton. It was the dominating science throughout the 19th century and is likely to have reached its zenith in the first half of our century with relativity and quantum mechanics, which were real revolutions. The analysis of the history of Physics and of these revolutions has been made by epistemologists such as Karl Popper or Thomas Kuhn. For Karl Popper a scientific theory cannot be proven but only falsified and this provides a criterion for a theory to be a scientific one. Thomas Kuhn, who was himself a physicist, saw science more as puzzle solving which ensures regular progress with, from time to time, the need for a revolution to modify the general framework in which this puzzle solving is done. The works of Karl Popper and Thomas Kuhn, although they are based essentially on an analysis of Physics, its history and its development, are usually considered as relevant to other fields of science and in particular to Biology. This is expressed for instance in the introduction written by Jacques Monod to the French translation of Karl Popper's major book *La logique de la découverte scientifique*.

Biology is a much more recent field of science than Physics. For centuries it was limited to a description or at best to a classification of living organisms. It is only with Darwin and Pasteur, in the second half of the 19th century, that was born a new Biology in which one tries to explain observations. The Darwin theory of evolution provides a coherent explanation for the diversity of living organisms. The discovery of microorganisms by Pasteur provided an explanation for fermentation and ended the concept of vitalism. Important consequences of Pasteur's discoveries were the belief that living organisms are made up of molecules similar to those already known by chemists and that the laws of Chemistry and Physics apply to these molecules inside the cell. This is at the origin of the belief that physicists should contribute to the understanding of life, expressed very well in Schrödinger's book *What is Life*. This book, although it is now more than 50 years old, remains stimulating and relevant. It was written at a period after the war when several physicists (Gamow, Max Delbruck and others) decided to work on biological problems which were, at that period, more free of ethical questions than work on atomic energy. Let us point out that in no place does Schrödinger

257

claim that Biology is reducible to Physics but he assumes that the physical laws apply to cell components and he analyses what are the resulting implications. One of these was that the material support of the genes must be a macromolecule.

It is a well known fact that constructive interactions between physicists and biologists have often been difficult to establish. This remains true although some physicists contributed in a very important manner to progress in Biology. The elucidation of DNA structure by Francis Crick and Jim Watson is a spectacular example of a fruitful cooperation between a physicist (at that time Francis Crick was still a physicist) and a biologist. The Phage group set up in California after the war, largely under the impulse of a distinguished physicist, Max Delbruck, is another example of successful collaboration between scientists of the two disciplines. But, in fact, Francis Crick very quickly became a biologist and his Physics' background was not very relevant in his later work. In his book of recollections, *What a Mad Pursuit*, he analyses the differences between Physics and Biology. He says for instance:

"The basic laws of Physics can usually be expressed in exact mathematical form, and they are probably the same throughout the universe. The "laws" of Biology, by contrast, are only broad generalizations, since they describe rather elaborate chemical mechanisms that natural selection has evolved over billions of years".

There are many other comments on relations between Physics and Biology in the book, but I selected this one which seems to me specially relevant as it includes some of the key words of the subject I am discussing: e.g. evolution, mathematical description.

The criteria to validate a scientific approach are very similar in Physics and Biology, but the methods that can be used are different for the two disciplines. I shall try to illustrate this point by a few examples. The finality of all biological objects is self reproduction. This is true for a virus, for a human being and all living organisms in between. This is not true for the molecules which constitute these organisms. The reproduction of an organism, however, requires that of its constituents. There is nothing of this sort in an object studied in Physics. So, the same molecule seen as a pure physical object is analysed in a completely different way when it is considered as an active part of a living cell. The DNA molecule is a striking example of this dichotomy. For a physicist it is a polymer, more precisely a polyelectrolyte, which has properties that are worthy of study. This interest arises in particular from the fact that chemists do not know how to make an artificial polymer as long and as monodisperse in length as the DNA molecules found in a cell or a virus. For a biologist, on the other hand, DNA is a molecule which carries genetic information and the relevant questions are: How is this information stored? How is it read? And, during cell division, how is it reproduced to be transmitted to the daughter cells? An ancillary question is: How is this very long molecule folded to fit into the nucleus of the cell? The physical properties of DNA that are determined by physicists are of limited relevance in answering these biological questions, in particular because, in the nucleus, DNA is always associated with specific proteins, which play key roles in the three points that the biologist wants to understand.

The DNA example illustrates well the point that in Physics one looks for properties which simply exist without a finality, whereas in Biology one must take into account biological finality, namely reproduction.

The concept of finality is meaningless in Physics. So, the nature of the scientific questions and the way of formulating them are different in Physics and Biology. In Physics one studies or one looks for molecules or material with specific properties (e.g. high electrical conductivity), perhaps useful for some application. In contrast, in Biology one observes and studies the properties of a molecule or an assembly of molecules and one tries to understand how these properties can be responsible for cell self replication and how they can account for evolution.

The DNA example also illustrates another difference of approach in the two fields. The physicist requires material that is as simple as possible. It is not by accident that DNA has been one of the very few biological objects studied by physicists. This is because it can be analysed as a simple polymer, although the non-repetitive occurrence of four bases in the sequence makes that it is not really a polymer in the usual sense of the term. But the approximation is good enough to use and the Physics of polymers and polyelectrolytes can be usefully applied to DNA. Physicists carefully avoid more complex biological objects and very little Physics has been done on proteins which, because of their tertiary and quaternary structures, cannot reasonably be assimilated to polymers. Physicists are interested in microtubules, however, for which they have the correct feeling that they are familiar objects because of their regular structure that correlates with a mechanical function. The biologist does not select the objects that he will study. He has to work on those which he finds in cells or organisms even if they are very complex.

The search for simplicity by the physicist leads him to an approach through *models*. The aim of the model approach is to find a system which is simpler than the real one, and consequently simpler to describe by mathematics, but close enough to the real one so that this description will be a good approximation to reality.

The word *model* is also used by biologists, but with a different meaning. In fact it is used in Biology within two contexts. The first is in naming the organism selected to study a specific function. For instance, in the first studies of developmental Biology many scientists worked on the fruit fly, *Drosophila*. This specific fly was their *model*. The motivation for the choice is to utilize an organism on which the biological experimental methods used (which we shall discuss later), in this case genetics, are easier to apply because of previous knowledge. But the fly is a real system, not really biologically simpler than a mouse or an elephant. In the second context, the word *model* is used in Biology to qualify the description of a hypothetical mechanism proposed to describe a function on the basis of experimental data. For example, the biologist will propose a model for DNA transcription. This use of the term *model* is rather similar to the expression "empirical theory" in Physics. So, both uses of the word *model* in Biology refer to something quite different from its meaning in Physics, where it involves mathematics and refers to a method central to research in Physics.

The importance of the model approach in Physics is a reason why the physicist can deal only with simple systems. Once again I must repeat that the biologist has no choice but to deal with very complex systems. The model approach, as utilized in Physics, does not apply to them. Some physicists, for example, have attempted to analyze the brain, modeling neurons as simple flip-flop units. The complexity of real neurons, involving a large number of synapses and neurotransmitters, is such that it is very doubtful that this physicists' approach will provide relevant results.

The complexity of biological systems is not only a fact that one must take into account, but also a necessity for life at all levels from within cells to organisms, at least as we know it from its manifestations on earth. To try to make this clear one must come back to the finality of a living object: reproduction. This is already complex in itself as replication of the cell requires a controlled replication of all its components. One knows that the information necessary for replication is stored in the genome, namely in a DNA molecule. But between this molecule whose replication is already rather complex, involving about a dozen different proteins (enzymes) and other cellular components, there are many metabolic pathways that must operate in synchrony. It is not sufficient for the cell to synthesize its components, they must also be put in or transported to the right place. And in a multicellular organism controlled cell differentiation is required. Moreover, all this must function in various environmental conditions. If something goes wrong in one of the many metabolic pathways, alternative pathways take over to allow cell survival. The existence of such alternatives, at least in several cases, has been demonstrated by "knock-out" experiments in which the destruction or neutralization of a gene considered as essential often has no effect on the phenotype.

In this short discussion on the complexity of biological systems it is worth mentioning that diseases are perturbations to the normal life of an organism, in which (with few exceptions) several factors are involved. This is even true for viral disease in which the same agent (a virus) can induce several different diseases or have no visible effects, depending on other factors, at least partly environmental, which are not always identified. For example, the Epstein–Barr virus is, in western countries, responsible for a minor disease, mononucleosis, whereas in eastern Africa it induces a specific lymphoma and in some parts of China (and a few other places) it is responsible for a larynx cancer. This is not due to a genetic determinant as demonstrated by the fact that the local specificity is lost by people going to live in another country. Similar local determinants also exist in cancer; the types of cancer are very different from country to country.

We have seen that the word *model* is used with a very different meaning by biologists and physicists. There are other examples of this language barrier between the two disciplines. The barrier is not in the technical language which simply requires from the physicist interested in Biology a small effort to assimilate a fairly rational biological vocabulary. It is rather in the concepts used. The concept of model was an example of a source of misunderstanding. The concept of "relation between structure and function" is another. It is a concept of great importance to interpret the structure of a biological macromolecule in terms of its biological consequences for the cell. Organization of the DNA molecule in two complementary strands facilitates its replication. This is a typical structure–function relation. This concept is central in the understanding of the biochemistry of the cell and it is a sort of extension of the concept of chemical function, itself often not well understood by a physicist. There are no equivalent concepts in modern Physics.

Experimental methods in Biology are often very different from those used in Physics. This is related to the complexity of biological systems. As it was explained by Regis Kelly at the beginning of his lectures, there are essentially three experimental methods available to the biologist: genetics, in vitro systems and drugs. These methods, whose names indicate their specificity for Biology, all assume the following general framework. The functioning of a cell and its replication is the consequence of interactions between molecules. These molecules are synthesized following instructions encoded in genes (essentially DNA molecules). Their interactions can be modified by drugs which are other molecules interacting with one or several of them. This universally accepted, general framework defines the science of Molecular Biology. In an in vitro system, one puts in a test tube several molecules whose interactions are assumed to have a consequence (in the simpler cases: the production of a new molecule). An example of such an experiment can be found in the self assembly of a virus from its components. In some rare cases it is possible to put in a test tube the components of a simple virus, namely RNA and a protein which is present in the native virus in many copies. After some time, in appropriate conditions of temperature, pH, and salt environment, particles will be recovered that seem to be undistinguishable, by physical and biological tests, from native viruses. Such experiments tell us that the self assembly of some viruses can be achieved only as a consequence of the chemical recognition between molecules, which itself results from their structure. I must mention that such an experiment works only for two or three viruses. Very often such an in vitro experiment is not possible as this simple chemical recognition is not sufficient to assemble the virus. For instance, in the case of poliovirus there is a step of maturation in which some of the viral proteins are cleaved after assembly by an enzyme whose gene is part of the viral genome.

The poliovirus example shows how nature has solved a biological problem, namely: how is it possible in the same cell to have the dissociation of the virus releasing its infectious nucleic acid, on one hand, and the assembly of the new virus from the molecules newly synthesized as a consequence of this infection, on the other? Clearly, the biological process is not reducible to a pure physical process, although all of the sequential events are normal chemical events fully in agreement with physical laws. The thermodynamics of the system is that of any physical system. But the sequence of events is not predictable from physical laws. This example shows the power and the limits of in vitro experiments, which are the easiest to understand for a physicist.

The interpretation of experiments is often more ambiguous in Biology than in Physics. This is a result of the complexity of the biological systems. I already mentioned the difficulties encountered in assigning a precise role to a gene or more precisely to a gene product. One finds a similar difficulty when using controlled mutagenesis to replace one particular amino acid by another in an enzyme. The fact that this replacement induces a reduction in the efficiency of the enzyme does not necessarily mean that this amino acid is directly involved in the interaction with the substrate. The effect may be due to a slight change in conformation of the enzyme that will affect its interaction with the substrate. These kinds of indirect effects are very misleading for the physicist who will often draw wrong conclusions too quickly. It is interesting to note that scientific articles in Biology very rarely use the expression "this experiment proves that ....." but rather "this experiment suggests that ...".

In Physics, theory plays a key role. At every stage of the development of the discipline the results are and have always been assembled and synthesized by general laws such as the law of universal attraction. Put together these laws make up a theory such as quantum mechanics or the theory of relativity. These laws are universal and have no exceptions. If an exception is found, the theory is revised. This is the way Physics progresses, as carefully described by Karl Popper. The situation is very different in Biology. There are very few general biological laws; and these nearly always have exceptions. When Francis Crick established the basis of Molecular Biology he put forward the "central dogma" which says that the flow of information in the cell goes from DNA to RNA and then to protein. One knows now that there are cases where the flow of information goes from RNA to DNA. And in the prevailing theory of the prion (which is not accepted by everybody) it is assumed that there can be a flow of information from protein to protein. The experimental observations in Biology, therefore, instead of being assembled in a rigid theoretical framework, are loosely assembled in something which is also named theory but which is far from having the rigidity of a physical theory, and which can, without major problem, accommodate exceptions. This has consequences for experimental work. Observation still plays a key role in Biology, with methods which can be very elaborate and very often rely on physical techniques. This is for instance the case with structure determinations of macromolecules by means of X-ray crystallography or NMR.

The universality of physical laws is so true that they are the same throughout the universe. If there exists a planet on which life exists, all physical laws found on earth will also be valid on this planet, but the type of life, defined as a self replicating system, could be based on very different principles and different chemistry than the ones prevailing on earth.

Altogether, Biology is a very concrete science in which abstraction plays a very limited role. This means that it is possible to teach Biology to physicists at the high level of problems dealt with by current research. The reverse is simply impossible. One knows that it is already difficult to teach solid state Physics to a nuclear physicist. The role of Mathematics is also, with some exceptions discussed below, limited in Biology. "Biology", according to Schrödinger, "can be explained without mathematics, not because it is simple enough for this, but because it is far too complex to be accessible to this science". There exists, nevertheless, a rich field of theoretical biology and there are cases where Mathematics is useful. The first one is found in genetics. Mendel's laws that describe the transmission of characters in heredity are expressed in mathematical terms which are indeed very simple, so simple that they can often be replaced by words. Let us note that these laws have exceptions, so are different from laws in Physics. A second case, which in fact derives from the first, is Biology of populations, namely the study of how a feature, a characteristic or a disease for instance, propagates in populations. Epidemiology can be considered as a special case of this discipline. Mathematics, in particular statistics, is necessary and widely used in this domain. Mathematics is also required to describe and explain oscillatory phenomena often observed in biological systems.

The role of time is very different in Biology and Physics. A biological system always evolves, as all cells (or nearly all of them) are always at some stage of their self replication and in the long term Darwinian selection is driving evolution. The contemporary ecosystem now includes species, man for instance, which were absent millions of years ago while others, such as dinosaurs, have disappeared. On still larger time scales, one must take into account that living systems have not always existed and the origin of life is itself an important scientific problem. In Physics most of the systems are stable. A piece of copper will remain identical to itself for extremely long periods and its physical properties, its electrical conductivity for instance, do not evolve. Radioactivity, which describes the change with time of a substance, was an unexpected phenomenon in Physics at the time of its discovery. Time appears also as a factor similar to the one in Biology in side branches of Physics like Geophysics, a domain obviously linked to the origin of life.

In conclusion, I have tried to show that the well recognized difficulties in the dialogue between biologists and physicists are related to rather fundamental problems. This does not mean that this dialogue is impossible. Political life shows that no dialogue is impossible if every one recognizes that there are differences and tries to understand others. An arrogant attitude by physicists leads nowhere. The same is true of a systematic rejection of the physical approach by biologists. But a physicist who understands that biological problems are different from those that he is accustomed to deal with and certainly not reducible to Physics may contribute, as Max Delbruck has done, for example, to important progress in Biology. The physicist needs not become a biologist, but he must understand what are the biological problems and their specificity. The organization of this Summer School could induce physicists to go in that direction as they have received lectures on some of the most important problems in modern Biology given by biologists actively pursuing first class research in these areas.

Textbooks which give a good comprehensive view of the two scientific fields are for Physics: *Lectures in Physics*, by R.P. Feynman, R.B. Leighton and M. Sands (California Institute of Technology, Addison Wesley, 1963), for Biology: *The Biology of the Cell*, by B. Alberts, D. Bray, J. Lewis, M. Raff, K. Roberts and James D. Watson (New York, 1989). The Biology text book is easily readable by a physicist. I am not sure that the reverse is true. It should be noted also that a textbook on Biology from 1963 would be completely obsolete. Most of Feynman book is still valid.

References

[1] F. Crick, *What a Mad Pursuit* (New York, 1988).
[2] H.F. Judson, *The Eighth Day of Creation* (New York, 1979).
[3] T.S. Kuhn, *The Structure of Scientific Revolution* (Chicago, 1962); Second edition enlarged (1970).
[4] T.S. Kuhn, *The Essential Tension* (Chicago, 1977).
[5] K.K. Popper, *The Logic of Scientific Discovery* (London, 1959). French translation with an introduction by Jacques Monod (Paris, 1973).
[6] E. Schrödinger, *What is Life? The Physical Aspect of the Living Cell* (Cambridge, 1967).